Climategate: The CRUtape Letters

Steven Mosher and Thomas Fuller

Copyright 2010, by Steven Mosher and Thomas Fuller

All rights reserved.
This book may not be reproduced, in whole or in part, including illustrations, in any form (beyond such copying as permitted by Sections 107 and 108 of the U.S. Copyright Law and except by reviewers for the public press) without written permission from the authors.

ISBN: 1450512437

EAN-13: 9781450512435

Steven Mosher offers this book to his son Logan Mosher, and Thomas Fuller to his wife Harivony.

And of course to the memory of C.S. Lewis, author of The Screwtape Letters, a delicious tale of temptation, resistance and renewal.

Acknowledgements

Writing a book in less than a month presents some challenges, and we would not have been able to meet them without the generous help of a large number of people. In particular we would like to point out the help offered by Martin Kokus, who edited this book in two days while being literally snowed under in Pennsylvania, Steven Lawrence, who did a brilliant job in producing the cover art and many of the graphs in this book in about 30 seconds flat and Dave Archibald, who kindly gave us permission to use two of his graphs.

We're also grateful to Steve McIntyre and Anthony Watts, for speaking with us, giving us permission to use material from their weblog and much more besides. And we're grateful to Lucia Liljegren just for existing.

For Steven Mosher, the posters at Climate Audit for educating me with special notice to: bender, UC, RomanM, Lucia, JeanS, and Willis Eschenbach; My friends for supporting and encouraging me: Charles Rotter, Jimmy Detels, Jayne Jones, Hutton Moffitt, Chris Pavis, Ron Preston, Sam Naughton and Ian King.

For Thomas Fuller, the commenters at Examiner.com, Ted Fuller Sr. and Zina Besirevic.

Obviously, a one-month-book is history in a hurry. And although we are grateful to everyone mentioned in this page, any errors in this book are ours to have and to hold.

PREFACE	7
Two Big, Fat Disclaimers	8
It's Worse In Context	9
Before the Beginning	11
What's At Stake	11
Cowboys and Indians	14
The Oldest Lie About Global Warming	14
CHAPTER ONE: SCIENCE MEETS THE INTERNET	16
Global Warming Scares Are Not Exactly New	16
This Time It Was Serious	16
Acronym Soup: The WMO and UNEP create the IPCC! Hurray!	17
Role of the IPCC:	18
Cast of Characters and Institutions	19
Meet the Team	19
The Blogosphere	23
Realclimate.org	23
ClimateAudit.org	25
Wattsupwiththat.com	27
The Warring Factions	30
CHAPTER TWO: THE DATA WAS NOT IN ORDER	32
More Than Mere Mortals Should Ever Be Forced To Learn About UHI	33
Sidewalks Get Hotter in the Summer Sun Than Grass	33
CHAPTER THREE: THE DEVIL AND MR. JONES	41
2005 FOIA: Jones's nightmare	49
Coming to Grips With Freedom of Information	51

CHAPTER FOUR: A PAPER IN PURGATORY 58
The Race against the clock 61
The Cat Bird Seat 64
Perception is often reality 71
CHAPTER FIVE. FREE THE DATA; FREE THE CODE 75
2007: The Year of UHI and FOIA 79
CHAPTER SIX: AN ARMY OF DAVIDS 97
Auditing the IPCC process 110
GISSTEMP: The Sacrificial Lamb 111
The End of the World 120
2009 122
CRU Refuses Data Once Again 123
CRU Excuses 124
CHAPTER SEVEN: HELL WEEK 128
CHAPTER EIGHT: THE CRUTAPE LETTERS 151
Hide the decline 152
Changing the questions we ask 163
Climategatekeeping 164
Possible Motives 173
Policy Impacts 175
CHAPTER NINE: AFTER THE GOLD RUSH 177
Real World Consequences of Climategate 178
The Real Crime in Climategate 179
Infant Science, Infantile Scientists 180
Global Warming and Cotton Candy 181
Global Warming and the Wolf 183

PREFACE

In late 2009, over 1,000 emails, attachments and files containing computer code were posted on an anonymous internet site. A few weblogs that focused on global warming received comments alerting them to the existence of these files. News of their existence quickly spread, and thousands of people downloaded the documents. This is the story of this event.

The emails and documents were communications between a small team of elite climate scientists and paleoclimatologists that had heavily influenced the IPCC's view of climate change. They had radically changed the IPCC's views in fact, and had almost convinced the world that temperatures had never been higher than they are today, and that they were climbing rapidly.

But the leaked files showed that The Team had done this by hiding how they presented data, and ruthlessly suppressing dissent by insuring that contrary papers were never published and that editors who didn't follow their party line were forced out of their position. And when Freedom of Information requests threatened to reveal their misbehavior, the emails showed them actively conspiring to delete emails to frustrate legitimate requests for information. Worst of all, one scientist threatened to actually delete climate data rather than turn it over—and that data is still missing.

We are writing this in December 2009, long before the story is over, and even before the implications can be truly evaluated. But given intense interest in the subject, we thought it would be useful to tell this story in order to deepen understanding of what happened, who were the principal actors, and why this took place. We note that discussion of Climategate, as the scandal has been dubbed, is full of defenders of these scientists who characterize the emails as a 'tempest in a teapot,' saying that 'boys will be boys' and that the science isn't affected. We don't think any of those statements are true.

As we write, the COP15 summit in Copenhagen has just wrangling over the commitment of hundreds of billions of dollars per year to developing countries to fight global warming. The U.S. Environmental Protection Agency has issued an endangerment finding declaring CO_2 a danger to public health.

The issue is worth $1 trillion a year, the amount that many environmentalists consider the appropriate sum to throw into the fight against global warming. With such astronomical sums at stake, getting the science right would seem to be at the heart of the discussion.

What this scandal (and hopefully this book) shows, however, is that for the scientists involved, even more important than getting the science right was getting the message perfect. And as the scandal plays out in the future, we think we can show that for many of the participants in this story, getting the message right meant ignoring holes in the science, shutting up those who disagreed and hiding the data from those who distrusted them.

This is not an easy book. It shows a sample of the emails chronologically arranged as we discuss separate topics. We discuss institutions, weblogs and science that we consider important. We realize this makes reading this a bit difficult. We've provided narrative summaries which we call 'Cheat Sheets' at the beginning of each chapter to try and help. But it's important to realize that the audit trail shows a pattern of improper (to say the least) behavior—and so we're asking our readers to do a little extra work. This is not a narrative—it is a case. If you stick with us, you'll understand at the very least why we thought it was important enough to approach in this way.

For example, it would be easy to show several emails discussing the deletion of files in advance of a Freedom of Information request. In fact, these emails were among the first to be discussed, first in weblogs and later in major media. If you as a reader have followed this issue at all, you have probably seen this email from Phil Jones:

> *And don't leave stuff lying around on anonymous download sites—you never know who is trawling them. McIntyre and McKitrick have been after the Climatic Research Unit ... data for years. If they ever hear there is a Freedom of Information Act now in the United Kingdom, I think I'll delete the file rather than send it to anyone.*

Using a timeline, however, allows us to construct The Team's long term strategy for avoiding the release of information that might prove embarrassing to them or damaging to the case they are trying to make through IPCC reports. As one of the defenses offered by supporters of The Team is that they may have acted rashly at times in writing emails, (with the follow-up of 'who hasn't?'), we think it more important to show that they decided on strategies of professional misconduct and pursued those strategies for a long period of time.

Two Big, Fat Disclaimers

We have taken sides in this analysis. Our critics will say that we took sides before we started, and although we are confident we have approached this objectively, there may be a little truth to that.

But—and it's a big but—although we are harsh in our criticism of the actions of this group of climate scientists and paleoclimatologists known as The Team, readers need to understand two things:

1. Our criticism does not extend to criticism of the theory of global warming. Both your authors believe global warming exists, is a problem and needs to be addressed. We just don't think it poses a catastrophic threat to civilization. We explain in detail below.

2. Our criticism should not be construed as criticism of the majority of scientists investigating our climate, its effects and possible changes to it in the future. We have communicated with a large number of climate scientists, and they are not at all like The Team in either attitude or behavior.

We are tough on the scientists we call The Team, and we think deservedly so. But we want to stress from the outset that we do not for one minute believe there is any evidence of a long-term conspiracy to defraud the public about global warming, by The Team or anyone else. What we find evidence of on a much smaller scale is a small group of scientists too close to each other, protecting themselves and their careers, and unintentionally having a dramatic, if unintended, effect on a global debate.

We take some comfort in the old story by C.S. Lewis, The Screwtape Letters, where Screwtape, the wise demon, advises Wormwood, the apprentice, on how to corrupt a British man's soul and secure for him a place in hell. Although Screwtape's advice is both sage and sordid, the plot fails and the British man ends up in heaven. We hope the CRUtape letters chronicled here eventually provide a similarly satisfactory ending.

It's Worse In Context

The response from the scientists involved in this controversy and their defenders is that critics have taken emails out of context that make their emails and behavior look worse than it actually is. That's one reason we're writing a book rather than a magazine article or a paper—to make sure that the context is there. For we believe that putting the emails into context shows that what happened is actually worse than what has been reported so far in the media.

But before we go through the timeline to establish this context, we should be specific about what we think the emails show. In this book we see scientists adopting the attitudes and behaviors of political activists and marketers:

- The scientists known as 'The Team' hid evidence that their presentation for politicians and policy makers was not as strong as they wanted to make it appear, downplaying the very real uncertainties present in climate reconstruction. The tree ring data was useful to them because it appeared to indicate that the most recent warming we have experienced was unprecedented and dramatic. But it inconveniently declined during the last few decades when they wanted it to increase the fastest; so they replaced the tree ring data with instrument data.

Michael Mann: *But that explanation certainly can't rectify why Keith's data, which has similar properties to Phil's data, differs in large part in exactly the opposite direction that Phil's does from ours. ...So, if we show Keith's line in this plot, we have to comment that "something else" is responsible for the discrepancies in this case.*

Phil Jones: *I've just completed Mike's Nature trick of adding in the real temperatures to each series for the last 20 years (i.e. from 1981 onwards) and from 1961 for Keith's to hide the decline.*

This happened despite the fact that many scientists they worked with had doubts about the material to be presented. One of the key figures in Climategate, Keith Briffa, goes so far as to say he believes something different than what their figures show:

I believe that the recent warmth was probably matched about 1000 years ago.

Malcolm Hughes writes, *I tried to imply in my e-mail, but will now say it directly, that although a direct carbon dioxide effect is still the best candidate to explain this effect, it is far from proven. In any case, the relevant point is that there is no meaningful correlation with local temperature.*

Ed Cook: *I have growing doubts about the validity and use of error estimates that are being applied to reconstructions.*

Tom Wigley: *I have just read the M&M stuff critcizing MBH. A lot of it seems valid to me. At the very least MBH is a very sloppy piece of work -- an opinion I have held for some time.*

Tom Wigley: A word of warning. I would be careful about using other, independent paleoclimatology ... work as supporting your work. I am attaching my version of a comparison of the bulk of these other results. Although these all show the "hockey stick" shape, the differences between them prior to 1850 make me very nervous. If I were on the greenhouse deniers' side, I would be inclined to focus on the wide range of paleoclimatology results and the differences between them as an argument for dismissing them all.

- The Team actively worked together to avoid compliance with the UK's Freedom of Information Act, alerting each other to possibly incriminating emails that needed to be deleted and advising each other on tactics to frustrate the intent of the FOIA. One scientist threatened to delete data rather than comply with the act, and data has in fact disappeared. At the same time, The Team was violating confidentiality agreements by sending the same data they refused to release to critics to their friends and supporters.

Phil Jones: *When the FOI requests began here, the FOI person said we had to abide by the requests. It took a couple of half-hour sessions—one at a computer screen, to convince them otherwise, showing them what Climate Audit was all about.*

Phil Jones: *If they ever hear there is a Freedom of Information Act now in the United Kingdom, I think I'll delete the file rather than send it to anyone.*

Phil Jones: *You can delete this attachment if you want. Keep this quiet also, but this is the person who is putting in Freedom Of Information requests for all the emails that Keith and Tim have written and received regarding Chapter 6 of the Intergovernmental Panel on Climate Change Report. We think we've found a way around this.*

Phil Jones: *Can you delete any emails you may have had with Keith regarding the latest Intergovernmental Panel on Climate Change report? Keith will do likewise*

- The Team actively worked to keep scientific papers that disagreed with their position out of the peer-reviewed literature. (They would then argue that skeptics should not be listened to because they did not publish papers in peer-reviewed literature.) They organized campaigns to replace editors of scientific journals who did publish skeptical papers, and also organized boycotts of journals that published contrary views. Perhaps most importantly, they violated principles of the peer-review process, which may serve to corrode trust in the methods scientists have used to communicate findings and improve understanding for over a century.

Michael Mann: *So what do we do about this? I think we have to stop considering Climate Research as a legitimate peer-reviewed journal. Perhaps we should encourage our colleagues in the climate research community to no longer submit to, or cite papers in, this journal. We would also need to consider what we tell or request of our more reasonable colleagues who currently sit on the editorial board...*

Phil Jones: *I will be emailing the journal to tell them I'm having nothing more to do with it until they rid themselves of this troublesome editor.*

Tom Wigley: *As you know, we suspect that there has been an abuse of the scientific review process at the journal editor level. The method is to choose reviewers who are sympathetic to the anti-greenhouse view. Recent papers in Geophysical Research Letters (including the McIntyre and McKitrick paper) have clearly not been reviewed by appropriate people.*

Michael Mann: *We probably need to take this directly to the Chief Editor at the Journal of Geophysical Research, asking that this not be handled by the editor who presided over the original paper, as this would represent a conflict of interest.*

And when someone tries to defend an editor,

Otto Kinne (Editor of Climate Research): *Chris de Freitas has done a good and correct job as editor*

The response is ruthless:

Michael Mann: *It seems to me that this "Kinne" character's words are disingenuous, and probably supports what de Freitas is trying to do. It seems clear we have to go above him. I think that the community should, as Mike Hulme has previously suggested in this eventuality, terminate its involvement with this journal at all levels—reviewing, editing, and submitting, and leave it to wither way into oblivion and disrepute*

Before the Beginning

This book tries to use a scandal to tell a wider story. It's very current—the Climategate scandal broke in November of 2009, and the broader story isn't even close to finished yet.

The scandal is about science, but also about politics and the media, and what some scientists are alleged to have done in promoting their point of view in the media to further their political point of view.

The wider story is about how we are going to look at the world going forward.

What's At Stake

Global warming has produced several ways of looking at the world. Here's one:

If you take the highest temperature of a day (TMax) and add the lowest temperature of that day (TMin), you get a number. Divide that number by two and you get an average for the day. If you add up the numbers for a month and plot them on a chart and compare it to the previous year, you get a better idea. If you add up all the numbers for a year and compare it to previous years, you get the chart below:

Fig. 1: Northern Hemisphere Temperatures for the last 1,000 years (The Hockey Stick Chart)
As constructed for MBH 1998

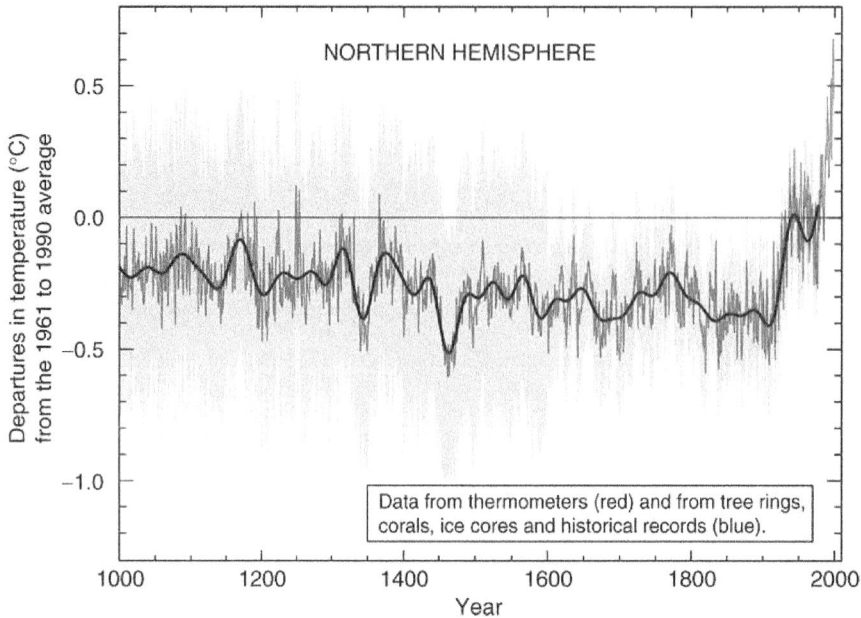

Steven Lawrence, after MBH 1998

Or do you? What if the temperatures are taken with different thermometers and thermocouples, and some of them are near a hot piece of asphalt and others have been moved to a busy airport? What if the first part of your series isn't from thermometers at all, but from estimates of temperature derived from analysis of the thickness of tree rings? And what if those estimates are suspect, declining during a period when we know temperatures rose?

The chart above was used by the Intergovernmental Panel on Climate Change in their 3rd report on the state of the Earth's climate, called TAR and published in 2001. Its dramatic shape and ominous red upward slope in recent times made it a perfect illustration of a particular point of view—that temperatures were rising faster than they ever had—that we were moving into uncharted and dangerous territory. That we needed to act now. But there are, as we will see, problems with that chart.

Here's another way of looking at the world—and it's the way we looked at temperatures for close on to 40 years, before worries about global warming made the previous chart so popular.

Fig. 2: Traditional View of Global Temperatures for the Past 11,000 Years

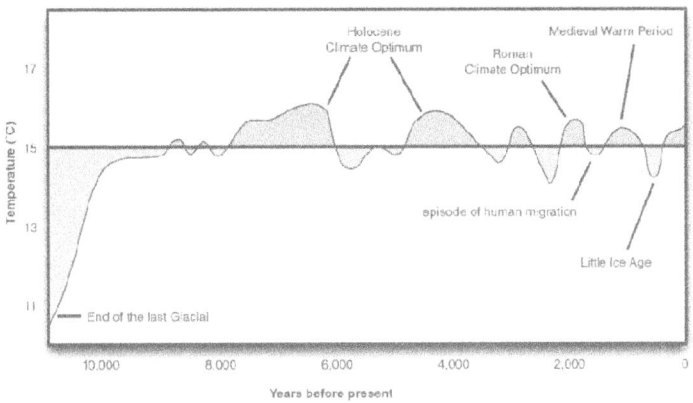

Reprinted by permission, David Archibald

In the figure above we see a rapid temperature rise in the first 1,000 years after the end of the last Ice Age and then fairly gentle fluctuations within a narrow band of temperatures after that, including the present day. According to this view of the world, what's happening to us now has happened before—without our emitting vast quantities of CO2. If this view is accurate, then worries about global warming seem, at best, premature, and the use of the term 'Optimums' to describe times when it was warmer than today may indicate that some warming is not all that bad.

There's yet another way of looking at the world and temperatures—over the very long term.

Figure 3: Temperatures and CO2 Concentrations in the Long View

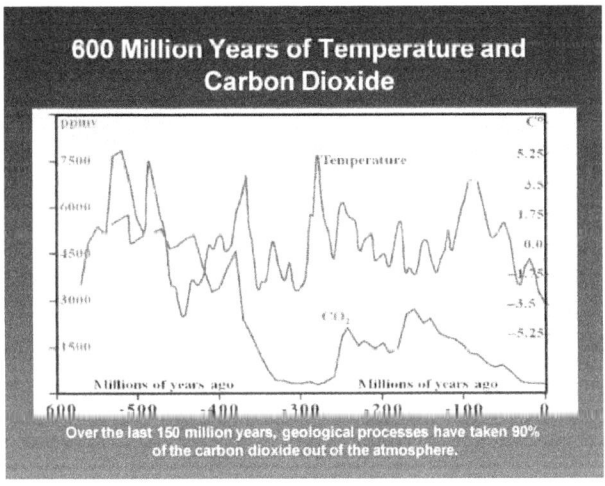

Reprinted by permission, David Archibald

Here we see that both temperatures and concentrations of CO2 are about as low as they have ever been. We also see here that there doesn't seem to be a simple or straightforward connection between CO2 concentrations and temperatures.

But the differences in these ways of looking at the world may have caused media savvy climate scientists to act in very unscientific ways to push their view of the world to the top and to 'disappear' competing views. This could be the first scandal caused by a PowerPoint chart. High crimes indeed.

Cowboys and Indians

Like global warming in general, this scandal offers many opportunities to sort players by personal preference or prejudice. In the United States, Republicans have tried to adopt, if not co-opt, the skeptic position while Democrats are generally champions of the environmental activists who see global warming as an immediate threat.

But it can also be portrayed as academic versus engineer—many of the most prominent skeptics and challengers of the status quo have had non-academic careers, while the most eminent activists have distinguished ivory tower credentials. The two groups have rubbed each other the wrong way for centuries, and they certainly didn't patch things up over global warming.

There is also an element of traditional communications vs. the internet here. Scientific journals and indeed academic communications move at a measured pace, with long periods between the publication of a paper and comments in response. This is markedly at odds with the less formal and more immediate give and take found throughout the internet. Those comfortable with one system are often irritated by the demands of the other—and it works in both directions.

Interestingly, both sides seem to sincerely feel that they are opposed by strong interest groups with an agenda that extends beyond the issue of climate change. For environmentalists, the skeptics are characterized as being funded by big oil and as tools of the conservative agenda. Skeptics look at the dispute as a well-funded assault on both reason and liberty. Readers should at least entertain the possibility that both sides are correct.

For many, the fight over global warming is just an extension of previous struggles, over other environmental causes or tobacco or geopolitics, and they came to this debate with agendas fixed and weapons ready.

But the Climategate scandal has the extra buzz of being partially a story about one man, Steve McIntyre, a Canadian mining engineer with a weblog, against the climate science orthodoxy. For onlookers, this created a storyline as old as stories themselves, the individual against the machine. Watching it all in real time on the internet was fascinating and addictive—and polarizing as well.

The Oldest Lie About Global Warming

You are about to enter a world where people in white lab coats don't play fair—not with facts, not with figures and certainly not with each other. Climategate is more than a scandal—it is the climate wars in microcosm. We'll give you an example that is both more than a century and less than a month old as evidence.

In 1895, Svante Arrhenius published a paper that described CO2 as a greenhouse gas and estimated the effect of a doubling of its concentrations. His calculations showed that worldwide temperatures would rise a hefty 6 degrees Celsius.

Al Gore, in his film An Inconvenient Truth, made much of this, saying it showed that claims of global warming were not new. But in 1906, Arrhenius had recalculated his figures, and published a new paper saying that doubling CO2 would only cause 2 degrees Celsius of warming. For some reason, Gore's film didn't pick up on that.

This embarrassing omission was picked up on by the skeptic community, and is one of the reasons that An Inconvenient Truth can't be shown in UK schools without a 'health warning' about films being used for political propaganda.

All very interesting, but the true nature of the debate about climate change was exemplified again in December 2009. John Rennie, writing in Scientific American (where he was editor in chief until very recently), did the exact same thing as Al Gore—citing Arrhenius' first prediction without mentioning his later, lower figure. As you will see in Climategate, untruths get recycled by both the great and the good—when they think no-one is watching them.

CHAPTER ONE: SCIENCE MEETS THE INTERNET

Global Warming Scares Are Not Exactly New

(From 'Fire and Ice, Business and Media') "It was five years before the turn of the century and major media were warning of disastrous climate change. Page six of The New York Times was headlined with the serious concerns of "geologists." Only the president at the time wasn't Bill Clinton; it was Grover Cleveland. And the Times wasn't warning about global warming – it was telling readers about the looming dangers of a new ice age. The year was 1895, and it was just one of four different time periods in the last 100 years when major print media predicted an impending climate crisis. Each prediction carried its own elements of doom, saying Canada could be "wiped out" or lower crop yields would mean "billions will die."

The same article, found on the website for Business and Media, talks of the New York Times headline reading, "Arctic Findings in Particular Support Theory of Rising Global Temperatures." But the date was Feb. 15, 1959. Glaciers were melting in Alaska and the "ice in the Arctic ocean is about half as thick as it was in the late nineteenth century." In 1953 William J. Baxter wrote the book "Today's Revolution in Weather!" on the warming climate. His studies showed "that the heat zone is moving northward and the winters are getting milder with less snowfall."

But it wasn't just warming that caused changes—cooling of 1 degree 'trimmed a week to 10 days from the growing seasons' between the 1940s and 1974, according to Time magazine.

In 1975 the popular media, supported by a handful of scientific papers, was regularly reporting on an upcoming Ice Age. But already, scientists were changing their focus to CO2 and its contribution to expected global warming. And lo and behold, this warming did indeed come to pass. Between 1975 and 1998 the earth warmed quickly and dramatically, as scientists, politicians, environmental organizations and large swathes of the general public began to view global warming as the challenge of our lifetimes.

This Time It Was Serious

In 1979 the World Meteorological Organization (WMO) convened the very first conference on climate change, the World Climate Conference. That conference concluded that "continued expansion of man's activities on earth might cause significant extended regional and even global changes of climate". In this conclusion we see the underlying premise that persists to this day: that mankind can change the climate. Within this frame of reference it becomes very difficult to suggest, study or publish contrary views—that the changes in climate are due solely to non-human causes or some combination of human and non-human causes. The conference also called for "global cooperation to explore the possible future course of global climate and to take this new understanding into account in planning for the future development of human society." And in this we see a premise that still holds in many quarters: global climate change requires global cooperation: mankind working together to save the planet.

Again, within this frame of reference it becomes difficult to suggest that countries may act locally in their own interest to prevent climate change in ways amenable to their citizens or to take actions to mitigate potential hazards. Finally, the conference issued a call to action and appealed to nations of the world "to foresee and to prevent potential man-made changes in climate that might be adverse to the well-being of humanity." The call to "foresee" and "prevent" has also shaped the debate seen over the past three decades, as the climate science community responded with tools, like General Circulation Models (GCMs) of the global climate, to "forecast" the climate of the future—and policy makers narrowed their focus to the "prevention" of climate change as opposed to strategies to adapt to climate change or mitigate the damages.

Over the course of the past 30 years this vision has culminated in an international effort to understand climate change, forecast the possible futures of the planet and recommend courses of action. In 1985 the WMO and others organized another conference and put carbon dioxide (C02) and other green house gases (GHGs) squarely in the crosshairs, concluding "as a result of the increasing greenhouse gases it is now believed that in the first half of the next century (21st century) a rise of global mean temperature could occur which is greater than in any man's history" This focus on man's changes to the atmosphere has shaped the science that follows in that other factors, such as land use changes, are nearly ruled out or given short shrift as causes of climate change. Ruled out as well is the possibility that our measurements of the climate are contaminated by land use changes, including what is known as the Urban Heat Island effect (UHI). This well known phenomenon is one of the focal points in the Climategate controversy.

In fact, Climategate as a story starts with the lengths one scientist, Phil Jones, (director of the Climate Research Unit and member of what we call The Team) went to in an effort to prevent others, principally Warwick Hughes, Steve McIntyre and Willis Eschenbach, from effectively reviewing his work and others on the Urban Heat Island effect.

In addition we see the claim made prior to any rigorous investigation that the expected climate change of the first half of the 21st century will be greater than any in man's history. This claim of unprecedented warmth and the science in support of that claim is also a focal point of the Climategate controversy. And again as the Climategate story unfolds we see two scientists, Keith Briffa and Michael Mann, working in conjunction with Jones to prevent McIntyre from reviewing their work on the climate of the past.

Acronym Soup: The WMO and UNEP create the IPCC! Hurray!

Two years later, in 1987, at the 10th Congress of the WMO, the stage was set for the creation of the IPCC, the Intergovernmental Panel on Climate Change. Together with UNEP, the United Nations Environment Program, it was decided that a new intergovernmental mechanism was needed, an organization focused on the scientific understanding of climate change and focused on actions that can be taken to prevent it. In 1988 the IPCC was born under the auspices of the WMO and UNEP.

The Role the IPCC set for itself is detailed in the 2004 bulletin. Probably the biggest misconception held by those who don't follow the global warming debate very closely is the idea that the IPCC does research themselves. They don't--they review established science and report on the results to policy makers every few years:

Role of the IPCC:

> "The role of the IPCC is to assess on a comprehensive, objective, open and transparent basis the scientific, technical and socio-economic information relevant to understanding the scientific basis of risk of human-induced climate change, its potential impacts and options for adaptation and mitigation. Review by experts and governments is an essential part of the IPCC process. The Panel does not conduct new research, monitor climate-related data or recommend policies. It is open to all member countries of WMO and UNEP."

The IPCC contracts out the review and editing process, and here's where the trouble actually starts. Some elements of climate change are really specialized. Not many scientists are involved in dendrochronology or paleoclimatic reconstruction of temperatures in the distant past. So the IPCC is more or less forced to contract with the scientists who wrote the papers to edit their own work and submit it to the IPCC and then write the appropriate section of the official reports. Even worse, the group is so small that they end up reviewing their own papers before they are originally published. Perhaps the central thesis of this book is that The Team, this group of scientists, didn't handle this situation well. We think the emails show they succumbed to 'groupthink' over a period of time, became defensive to any criticism, and resorted to increasingly desperate measures to protect their initial hypotheses. We are told, by scientists, that this is not rare in science. However, with global warming the stakes have gotten so large that academic infighting now potentially affects the world.

All aspects of their role come under scrutiny in Climategate. What the internal mails reveal is a process that is not open or transparent. Since the publication of AR4 (the Fourth Assessment published in 2007) individuals both inside and outside the IPCC process have had to fight and take legal measures to publish the proceedings of these expert reviews. What the emails reveal is a process driven by a select group of individuals who go to extremes to crowbar certain scientific studies into the review process while at the same time conspiring to keep others out.

The IPCC is charged with issuing reports on the state of climate science, an assessment of our understanding of the climate and man's role in changes to it: The IPCC reports started in 1990 with the first assessment report (FAR), followed in 1995 with the second assessment (SAR), and in 2001 with the third assessment (TAR) and finally in 2007 the fourth assessment report or AR4 was published. AR5 or the fifth assessment is due in 2013.

Almost at the same time that AR4 was published in 2007, the scientists involved in this controversy joined other climate scientists in saying the report was too conservative and that global warming would be faster and more destructive than depicted in AR4. This is strange, as many of these same scientists were heavily involved in the preparation and writing of AR4. But knowing that they had six years before the publication of the next report, they could publish what they pleased in the meantime to help shape political and public opinion.

Cast of Characters and Institutions

Meet the Team

As we talk throughout the book about The Team, we should at least introduce them.

The original 'Hockey Team' was named by Real Climate back in 2005, in the early days of triumph after the publication and subsequent popularity of the Hockey Stick Chart. This early team consisted of Michael Mann, R.S. Bradley and M.K. Hughes. They have since disavowed the name.

The Team as it is today still includes Michael Mann, Bradley and Hughes, but others have come on board. They include Keith Briffa, Phil Jones, Gordon Jacoby, Schweingruber, Rutherford, Crowley, Cook, Osborn and perhaps others who float in and out. For the purposes of our story, the principal characters are Michael Mann, Phil Jones and Keith Briffa.

Michael E. Mann, according to his biography on the Real Climate site, is a renowned climate scientist and author of more than 80 journal publications. His specialty is paleoclimatology, reconstructing the climate of the past by using "proxy" data such as tree rings, ice cores and sediments. He is most famous for a graph that has come to be known as the "hockey stick graph," a graph that assumed iconic status for believers in global warming, which ostensibly shows that the current warming we see is unprecedented in human history. In short, his work shores up the claims made in 1987 at the WMO conference. As of this writing his employer, Pennsylvania State University is investigating his role in the Climategate controversy.

Phillip Jones (born 1952) is the director of CRU and, according to his biography on CRU's website, is a professor in the School of Environmental Sciences at East Anglia in Norwich. His research interests include instrumental climate change and paleoclimatology, the study of historical and pre historical climate. His papers on the Urban Heat Island (UHI) from the 1980's to present play a vital role, not only in the IPCC understanding of the climate, but also in the Climategate controversy. His work in paleoclimatology, specifically work with Michael Mann and Keith Briffa is also central to the scandal. It's no understatement to say that almost every major paper written about the issue of the Urban Heat Island and every reconstruction of the climate of the past has relied on the foundational work of Phillip Jones. It's a cornerstone of climate science. Jones is also a key contributor to many of the publications of the IPCC.

The emails of Climategate also reveal that Jones wanted to play an active role in shaping the policy of the IPCC with regards to disclosing the data behind the science. That is, he wanted his experience in dealing with critics' requests for data to shape and inform IPCC policies. In the following email Jones discusses the next assessment report (AR5) of the IPCC. In this time frame, late July 2009, Jones and the climate scientists who worked on AR4 had just been through Freedom of Information requests pertaining to Chapter 6 of AR4, a chapter on the climate of the past.

Mail from Phil Jones to Tom Peterson of NCDC (1249045162.txt)

On something positive - attached is the outlines for the proposed [Chapters] in AR5/ WG1 [Working Group 1]. ... I have got the IPCC Secretariat and Thomas to raise the FOI [Freedom of Information] issues with the full IPCC Plenary, which meets in Bali in

19

September or October. Thomas is fully aware of all the issues we've had here wrt [with regard to] Ch 6 last time, and others in the US have had.

Cheers Phil

Jones is a member of the standing committee on data archiving and accessibility for the US Department of Commerce's NOAA. This appointment allows Jones to advise a division of the US government on the issue of archiving and data access. The mails will certainly raise questions about his ability to serve the public interest when it comes to the issues of data archiving and data access. Under his direction, East Anglia University's Climate Research Unit lost vital data, protected data at considerable cost to his organization, and Jones personally threatened to delete data if the law required him to turn it over.

In addition to compiling the global temperature index the staff at CRU are dedicated to understanding the climate of the past. One claim underlying the founding of the IPCC, it should be recalled, is that the climate change we are seeing is unprecedented in human history. Supporting that claim is the work of paleoclimatologist Keith Briffa. Together with Jones, Briffa sought to paint a picture of the climate of the past that removed any doubts we may have about the unusual changes we see in the climate today. Briffa's work and the work of other paleoclimatologists is covered in Chapter 6 of the AR4, referenced in Jones' mail above.

Keith Briffa, according to his biography on CRU's website, has been at the University of East Anglia since 1977, where he specializes in studying the climate of the past. His particular area of expertise is in using tree ring data to reconstruct the climate of the past. This field of science, known as dendroclimatology uses a variety of tree ring data, ring widths, wood density and isotopes to estimate or reconstruct the climate of the past. Those reconstructions play a role in supporting some of the founding claims of the IPCC, namely that the change we see today in the climate has not been seen before in human history, the implication of course being that most of the change can be attributed to human activity. One critical part of these reconstructions is the present day temperature record. This record, the data set created by Jones, is the lynchpin of many climate reconstructions. Without reliable data about the state of the climate in recorded history, reconstructing the past becomes impossible. Briffa's work surfaces in the Climategate files primarily because of his work on AR4 and in particular his work on climate reconstructions using tree rings. Briffa's data and methods employed by Jones have become the focus of the independent investigators.

In addition to CRU and the University of East Anglia one other organization in the UK plays a role in Climategate: The MET Office Hadley Centre for Climate Change. According to its website MET is a government office that receives funding from the Department for Environment, Food and Rural affairs (DEFRA), other government departments and the European Commission, the Hadley Centre has the following charter: They aim to understand the processes of the climate system, develop computer models or GCMs to simulate the climate and use those models to predict climate changes in the years to come. Finally, the center is tasked with monitoring climate change. In this role the Hadley Centre publishes the global temperature index created by CRU and Jones and this made them a target for Freedom of Information requests.

Throughout the world various other organizations work in support of the goals of the IPCC. The key organization at the center of the controversy is the Climate Research Unit or CRU. CRU is a component of East Anglia University in England (EAU) and is recognized as one of the world's leading institutions studying natural and human induced or anthropogenic climate change. According to its web site, its staff of around thirty research scientists and students has developed and published a number of the data sets widely used in climate research, most importantly the global temperature index, or CRUTEMP. The data and methods surrounding this critical piece of evidence of climate change is the principle focus of Climategate. The quest to independently investigate that data set in motion all that follows. The principal actor here is the director of CRU, Phil Jones.

Across the ocean there are also several American institutions that play a role in Climategate. They include:

- **NOAA**: National Oceanic and Atmospheric Administration, part of the U.S. Dept. of Commerce

- **NCDC**: National Climatic Data Center, world's largest archive of weather data, and the organization that maintains the core climate database, the **GHCN** (Global Historical Climatology Network)

- **NCAR**: National Center for Atmospheric Research, sponsored by the National Science Foundation

- **GISS**: Goddard Institute for Space Studies, part of NASA, conducts research to predict atmospheric and climate changes

The role of these institutions tends to be relatively minor although key individuals at each institution are involved. NOAA, the National Oceanic and Atmospheric Administration, is a scientific agency reporting to the Department of Commerce. As a part of their charter they conduct research on the climate, collect climate data and participate in the IPCC process. The key individuals are:

- **Tom Wigley**, former director at CRU, senior scientist at NCAR, and now working for UCAR (University Corporation for Atmospheric Research)

- **Thomas Peterson**, research meteorologist at NOAA, lead author for IPCC AR4

- **Tom Karl**, director of NOAA's NCDC, and lead author of sections of three previous IPCC reports

- **Jonathan Overpeck**: Professor of Geosciences at the University of Arizona and Director of the Institute for the Study of the Planet Earth

- **Susan Solomon**: Senior scientist at NOAA

- **Benjamin Santer**: Climate researcher at Lawrence Livermore National Laboratory

Working hand in hand with NOAA is the National Climatic Data Center in Ashville North Carolina. The NCDC, according to their website, archives 99 percent of all NOAA data, including 320 million paper records and over 1.2 petabytes of digital data. Among all the institutions involved, the NCDC has the best track record of providing data to all who seek it. The core database in question in Climategate is the NCDC's Global Historical Climatology Network or GHCN. GHCN is a collection of climate data from the US and the rest of the world. There are roughly 6,000 temperature stations represented in its database. This data is freely available online. Over the course of history several organizations around the world have contributed data to the GHCN, but it is important to note that some country data is not released to the GHCN or it is released under agreements that preclude it from being released to third parties. That raw unprocessed data is at the heart of Climategate. Jones wanted to protect it and his critics wanted access to it. While NCDC prides itself on providing data to researchers, there is clearly some tension within the institution. This mail between Tom Peterson and Phil Jones is evidence:

From: Thomas.C.Peterson@xxxxxxxxx.xxx   Date: Wednesday, July 29, 2009 12:07 pm   Subject: Re: This and that

Hi, Phil,

Yes, Friday-Saturday I noticed that ClimateFraudit had renewed their interest in you. I was thinking about sending an email of sympathy, but I was busy preparing for a quick trip to Hawaii - I left Monday morning and flew out Tuesday evening and am now in the Houston airport on my way home. Data that we can't release is a tricky thing here at NCDC. Periodically, Tom Karl will twist my arm to release data that would violate agreements and therefore hurt us in the long run, so I would prefer that you don't specifically cite me or NCDC in this. But I can give you a good alternative. You can point to the Peterson-Manton article on regional climate change workshops. All those workshops resulted in data being provided to the author of the peer-reviewed paper with a strict promise that none of the data would be released. So far as far as I know, we have all lived up to that agreement - myself with the Caribbean data (so that is one example of data I have that are not released by NCDC), Lucie and Malcolm for South America, Enric for Central America, Xuebin for Middle Eastern data, Albert for south/central Asian data, John Caesar for SE Asia, Enric again for central Africa, etc. The point being that such agreements are common and are the only way that we have access to quantitative insights into climate change in many parts of the world. Many countries don't mind the release of derived products such as your gridded field or Xuebin's ETCCDI indices, but very much object to the release of actual data (which they might sell to potential users). Does that help?

This email details a conflict at the heart of Climategate. Some countries make money from selling their temperature data to users. Scientists who acquire this data are then torn between their scientific principles, which argue that data must be shared to be verified, and their institutional obligations, which require them to abide by agreements. We also see that within the institutions different scientists view these data release issues with different attitudes. Most importantly, however, NCDC wants to stay out of the news.

The last two organizations, NCAR and GISS, play a slightly more tangential role in Climategate. NCAR or the National Center for Atmospheric Research is home to some of the scientists involved in the emails uncovered and GISS, the Goddard Institute for Space Studies run by noted scientist James Hansen is also home to Gavin Schmidt, a renowned climate scientist who runs the Blog "Real Climate." Dr. Schmidt is a climate modeler at NASA GISS and he specializes in modeling the past, present and future. While Schmidt is a principal and vocal defender of climate science, his role in Climategate is relatively minor and he is not implicated in any way.

The Blogosphere

As the IPCC was coming into being during the 90s, the Internet was blossoming, and in the mid 90's the print genre of the diary or personal journal was transformed into an electronic version known as a "weblog" which quickly got shortened to 'blog.' In many ways the story of Climategate illustrates the power of the Internet, an open and transparent medium that people can use to communicate with others. Whereas the IPCC is a centralized, rule governed body with policies and procedures, with aims and missions, with guidelines for publications and timelines for publication, and with budgets, the blogs are loosely organized, affiliated by common interests and readers, with varied emergent rules for conduct, with individualistic charters, with no guidelines for publication and very often no budgets. In the same way internet journalism challenged print journalism and mainstream media, the way science enthusiasts analyzed and reviewed events and publications on blogs challenged the way science and scientists communicated.

One of the ongoing struggles made clear by the leaked emails is that between the command and control, top-down dissemination of science through traditional channels, and the more chaotic and immediate broadcasting of information through informal channels on the internet. When history provides a broader perspective on the events covered in this book, this conflict will probably receive more attention than we give it here.

Realclimate.org

Not all blogs, however, are personal diaries or the efforts of a single individual. In the global warming debate the unofficial voice of the IPCC is the blog "Real Climate" or RC for short. Although in his mail to fellow climate scientists Schmidt denies that RC provides an "official position" on climate matters it is the de facto "voice" of climate scientists: The following mail sent by Gavin Schmidt details the positioning of RealClimate:

> *Subject: RealClimate.org Date: 10 Dec 2004 08:56:42 -0500 Cc: Mike Mann <mann@xxxxxxxxx.xxx>, Eric Steig <steig@xxxxxxxxx.xxx>, Ammannn@xxxxxxxxx.xxx, rbradley@xxxxxxxxx.xxx, aclement@xxxxxxxxx.xxx, rasmus.benestad@xxxxxxxxx.xxx, rahmstorf@xxxxxxxxx.xxx*

> *Colleagues, No doubt some of you share our frustration with the current state of media reporting on the climate change issue. Far too often we see agenda-driven "commentary" on the Internet and in the opinion columns of newspapers crowding out careful analysis. Many of us work hard on educating the public and journalists through lectures, interviews and letters to the editor, but this is often a thankless task. In order to be a little bit more pro-active, a group of us (see below) have recently got together to build a new 'climate blog' website: RealClimate.org which will be launched over the next few days at: http://www.realclimate.org The idea is that we working climate scientists should have a place where we can mount a rapid response to supposedly 'bombshell' papers that are doing the rounds and give more context to climate related stories or events. Some examples that we have already posted relate to combating dis-*

information regarding certain proxy reconstructions and supposed 'refutations' of the science used in Arctic Climate Impact Assessment. We have also posted more educational pieces relating to the interpretation of the ice core GHG records or the reason why the stratosphere is cooling. We are keeping the content strictly scientific, though at an accessible level. The blog format allows us to update postings frequently and clearly as new studies come along as well as maintaining a library of useful information (tutorials, FAQs, a glossary etc.) and past discussions. The site will be moderated to maintain a high signal-to-noise ratio. We hope that you will find this a useful resource for your own outreach efforts. For those more inclined to join the fray, we extend an open invitation to participate, for instance, as an occasional guest contributor of commentaries in your specific domain, as a more regular contributor of more general pieces, or simply as a critical reader. Every time you explain a basic point of your science to a journalist covering a breaking story, think about sharing your explanation with wider community. Real Climate will hopefully make that easier. You can contact us personally or at contrib@xxxxxxxxx.xxx for more information. This is a strictly volunteer/spare time/personal capacity project and obviously nothing we say there reflects any kind of 'official' position. We welcome any comments, criticisms or suggestions you may have, even if it is just to tell us to stop wasting our time! (hopefully not though). Thanks,

Gavin Schmidt

We see the frustration the scientists felt in getting the media and the public to understand the science as well as the desire to fight a rapid reaction battle against a host of critics who used the Internet to post their "refutations" of accepted science. This is a common theme that runs throughout the Climategate story. Working climate scientists publish their findings in the peer-reviewed journals of science: Nature, Science, and others. The process of publishing is long and tedious and open to manipulation as we shall see.

On the other side, often combating them, are a variety of voices from outside this mainstream publishing world. It's an internet world where anyone can post their ideas immediately, without review or rather with a wide-open unfettered review by hundreds or thousands or millions of readers. The culture shock is evident, and not just in Schmidt's email. A process that formerly took months to get a polite reply from a colleague was being challenged by a world where the scientist would be confronted by dozens, if not hundreds, of blog posts from people ranging from scientifically qualified to people who had never opened a book before picking up a pen. Scientists who might "flame" another scientist in private mails were exposed to an internet world where the "flaming" was out in the open.

Unlike the other blogs in the Climategate story, Real Climate (RC) has connections with a public relations organization, Environmental Media Services (EMS) and while that organization claims no control over the content, the mere fact that this politically progressive organization has anything to do with RC has caused skeptics to question whether RC lives up to its stated goal of discussing climate science objectively. In 2005, however, Scientific American recognized RC for its excellence in writing: "A refreshing antidote to the political and economic slants that commonly color and distort news coverage of topics like the greenhouse effect, air quality, natural disasters and global warming, Real Climate is a focused, objective blog written by scientists for a brainy community that likes its climate commentary served hot. Always precise and timely, the site's resident meteorologists, geoscientists and oceanographers sound off on all news climatological, from tropical glacial retreat to "doubts about the advent of spring."

The principal contributor is Gavin Schmidt, a scientist who works for NASA. However, the masthead at Real Climate includes several notable climate scientists, chief among them Michael Mann, both author and recipient of many of the Climategate mails. Other contributors include Eric Steig and Stephan Rahmstorff, who we will meet later.

ClimateAudit.org

The central (although not the most highly trafficked) player in the blogosphere opposing Real Climate is Stephen McIntyre's blog, Climate Audit. (CA) The principal focus of McIntyre's blog is "auditing" the work of climate scientists.

McIntyre's Wikipedia page notes, "McIntyre attended the University of Toronto Schools, a university-preparatory school in Toronto, finishing first in the national high school mathematics competition of 1965. He went on to study mathematics at the University of Toronto and graduated with a bachelor's of science degree in 1969. McIntyre then obtained a Commonwealth Scholarship to read philosophy, politics and economics at Corpus Christi College, Oxford, graduating in 1971. Although he was offered a graduate scholarship, McIntyre decided not to pursue studies in mathematical economics at the Massachusetts Institute of Technology.

McIntyre worked for 30 years in the mineral business, the last part of these in the hard-rock mineral exploration as an officer or director of several public mineral exploration companies. He has also been a policy analyst at both the governments of Ontario and of Canada. He was the president and founder of Northwest Exploration Company Limited and a director of its parent company, Northwest Explorations Inc.

As McIntyre explained when interviewed for this book, he has not taken a paycheck from any company in the past 23 years. Financially, he stands to gain in an economy that moves away from petroleum products. Economies based on electricity drive demands in the metal markets which in turn benefits the kind of mining interests that he is occasionally involved with.

At Climate Audit, McIntyre concentrates on two issues: Official records of past temperature, such as those produced by Phil Jones and CRU, and historical climate reconstructions, such as those produced by Michael Mann and Keith Briffa. The "audit" metaphor is an apt one for McIntyre's style and focus. McIntyre's interest in climate science began when advocates of the Kyoto Protocol used the now famous "hockey stick graph" from Michael Mann 1998 paper.

Faced with a journal publication from Mann or Briffa for example, McIntyre's first response is to get the data and methods from the author so that he can check their work for errors, whether they be minor errors, typographical errors or substantial errors. The task he attempts to perform is a basic quality check on the science. The Climate Audit blog started in 2005 after the founding of RC according to McIntyre he started the blog because of attacks on his work published at RC. In addition, as noted in the founding mail sent out by Gavin Schmidt, RC has a policy of "moderating" comments. How that process works is described here:

From: "Michael E. Mann" <mann@meteo.psu.edu>

To: Tim Osborn <t.osborn@uea.ac.uk>, Keith Briffa <k.briffa@uea.ac.uk>

Subject: update

Date: Thu, 09 Feb 2006 16:51:53 -0500

Reply-to: mann@psu.edu

Cc: Gavin Schmidt <gschmidt@giss.nasa.gov>

guys, I see that Science has already gone online w/ the new issue, so we put up the RC post. By now, you've probably read that nasty McIntyre thing. Apparently, he violated the embargo on his website (I don't go there personally, but so I'm informed). Anyway, I wanted you guys to know that you're free to use RC in any way you think would be helpful. Gavin and I are going to be careful about what comments we screen through, and we'll be very careful to answer any questions that come up to any extent we can. On the other hand, you might want to visit the thread and post replies yourself. We can hold comments up in the queue and contact you about whether or not you think they should be screened through or not, and if so, any comments you'd like us to include. You're also welcome to do a follow-up guest post, etc. think of RC as a resource that is at your disposal to combat any disinformation put forward by the McIntyres of the world. Just let us know. We'll use our best discretion to make sure the skeptics don't' get to use the RC comments as a megaphone...

mike

That means that RC reserves the right to publish comments from readers or not, a right they exercise frequently. McIntyre complained that RC has blocked his comments. That muzzling played a role in his decision to start his own megaphone, Climate Audit. In 2007 his blog won the Best Science Blog of the Year award, illustrating the futility of muzzling speech. RC has yet to win the award, despite copious (and often valuable) information about climate science maintained on the site.

The function McIntyre provides essentially replicates the number-checking done internally at many organizations, but which seems to be missing from climate science. On the surface McIntyre seems an unlikely candidate to audit climate science. However, McIntyre's career has exposed him to a rigorous due diligence process that relies on the collection of data and the statistical analysis of data. The math skills required for that process are considerable and McIntyre was suitably prepared. Winner of the Ontario high school math contest in 1965, he went to the University of Toronto, where his focus was pure mathematics, algebraic topology, group theory, and differential manifolds. He was offered a scholarship to study mathematical economics at MIT but instead studied philosophy, politics and economics at Oxford.

Climate Audit has several contributors in addition to Stephen McIntyre. These authors include retired statistics professors and specialists in time series analysis. At times the blog posts are only comprehensible to those with advanced degrees in mathematics or statistics. Perhaps first among these co-authors is Ross McKitrick, who co authored a paper with McIntyre in 2005 on Mann's "Hockey Stick". According to the University of Guelph website, McKitrick holds a PhD in Economics and is a professor at the University of Guelph. His area of specialty is environmental economics and statistical methods in paleoclimatology.

McKitrick is a lightning rod for The Team's defenders, as his past connections include work at organizations they despise. (For some environmental activists, a one-time association with the wrong company or think tank taints you forever, and they will refuse to even listen to you based on your history. We write more about this below.) Our reading of McKitrick's work shows no discernible political bent or bias, but readers are urged to form their own opinions.

Both McIntyre and McKitrick share a common attitude toward the principles of disclosure with respect to the data and methods they use in their publications. Taking their lead from requirements in econometric journals, both stress the vital importance of backing up their claims in published work with the actual data used in their calculations and the computer codes required to perform the calculations. As a result, posts or articles at CA come complete with the data required to replicate the results shown and the computer code to generate those results.

This approach allows all readers to see the results for themselves, to check McIntyre's work and the work of others. With many eyes looking at the same problem and many eyes looking for biases on the part of the investigators errors are quickly surfaced and dealt with. This approach to open data and open source computer code stands is actually a subplot for the entire Climategate story. From his entry into climate science McIntyre has focused on getting access to the data and methods of key climate scientists: Phil Jones' temperature data and code and paleoclimatologists' data and code. His approach to getting access to the data and code is as follows. He writes to the scientists publishing the results and asks them for the data, and follows up requests that are denied with a request to the journal that published the paper. Unlike the field of econometrics, the journals that publish climate science do not uniformly require that authors supply the data and code to back up their claims. If neither the scientists nor the journals are forthcoming, McIntyre has availed himself of legal methods of getting the information he desires, namely Freedom of Information requests.

Wattsupwiththat.com

The most popular blog in the Climategate story is a skeptically oriented news and opinion blog called Watts Up With That (WUWT). Winner of the 2008 Science Blog of the Year award, WUWT was started by television meteorologist Anthony Watts. At its inception in 2007, WUWT was focused primarily on a volunteer project organized by Watts. That project, surfacestations.org, had one goal: document the state and condition of every station in the US used by climate scientists to record temperature and examine the problem of UHI, the Urban Heat Island effect. This project is largely complete, with volunteers from around the country scouring the land to provide photo documentation of these surface stations.

Like McIntyre, Watts has an eye for detailed records and a "show me the data" attitude. One thread of the Climategate story winds around the issue of the temperature data used by Jones in his global temperature series. That temperature is collected from surface stations on a daily basis. After visiting a few stations in California Watts was shocked by what he discovered: the stations which were supposed to be located in areas free from UHI were in fact located in areas where their readings could be influenced by non climatic factors. Stations which were supposed to be located away from buildings and trees were found next to trash burning bins, on rooftops, under trees. According to Watts the implication was clear: the temperature records of climate science were not measuring changes in the climate; they might very well be measuring changes in the build-up of human communities around these stations instead. The record, which Jones had said was free from UHI, may, in fact, be infected by it. Jones and the climate scientists had assumed that stations were properly maintained and properly situated away from human influence. Watts put this supposition to the test and his volunteers visited well over 1,000 stations in the USA, photo documenting each one:

Figure 4: Temperature Measuring Station, 2009

Reprinted by permission, Watt's Up With That

The communities that read WUWT and CA overlap somewhat, although WUWT is much more 'general interest' and CA is far more technical. At the startup of his blog, Watts was a frequent contributor at CA and his work was followed closely by Steve McIntyre. To this day the two blogs help each other out and when CA was crushed by the traffic caused by the Climategate scandal, Watts was there to help McIntyre out by hosting his articles. However, the focus of the two blogs is very different. Posts at CA are tightly focused on published science and particular issues in climate science: the temperature record and paleoclimatology. McIntyre has little time to post or comment on science from the fringe or unpublished science. He likewise heavily discourages discussions of policy. WUWT, on the other hand, covers whatever is new and interesting in climate studies and politics. If Real Climate is the unofficial voice of accepted climate science, WUWT is the soapbox for those who are skeptical of climate science.

There are two other blogs that play a minor role in Climategate. The first is Jeff Id's "Airvent." The second is Lucia Liljegren's Blackboard, AKA "Rank Exploits." The owner of the Airvent is anonymous. A self-described aeronautical engineer, the principal contributor is a frequent commenter on CA and WUWT and occasionally posts on RC. Like McIntyre, "Jeff Id" is entirely focused on replicating the work of climate scientists, in particular paleoclimatologists. When the person who "hacked" the CRU systems and took possession of the mails looked for a place to announce their existence, he placed a link to the file in a comment on Airvent, as well as two other blogs. The Blackboard is a blog run by Lucia Liljegren. Lucia, a self described "Lukewarmer," appeared first as a commenter on CA. With a degree in mechanical engineering she held her own in debates on CA with working and retired scientists. Her blog does not receive the traffic that CA or WUWT or RC does, but she has a unique perspective and on occasion Gavin Schmidt, principal contributor at Real Climate, will show up to comment on her work. In a style only seen on the Internet Lucia combines discussions of differential equations, statistics, haiku, knitting and baked goods into a delightful and civilized discussion. Her blog is perhaps the one place where the warring factions come together to discuss issues. When the Climategate story broke, it broke in a comment on her Blog:

> Steven Mosher (Comment#23722) November 19th, 2009 at 1:55 pm
>
> Lucia,
>
> Found this on JeffIds site.
>
> http://noconsensus.wordpress.c.....en-letter/
>
> It contains over 1000 mails. IF TRUE …
>
> 1 mail from you and the correspondence that follows.
>
> And, you get to see somebody with the name of Phil Jones say that he would rather destroy the CRU data than release it to McIntyre.
>
> And lots lots more. including how to obstruct or evade FOIA requests. and guess who funded the collection of cores at Yamal.. and transferred money into a personal account in Russia
>
> And you get to see what they really say behind the curtain.. you get to see how they "shape" the news, how they struggled between telling the truth and making policy makers happy.
>
> you get to see what they say about Idso and pat Michaels, you get to read how they want to take us out into a dark alley, it's stunning all very stunning. You get to watch somebody named Phil Jones say that John Daly's death is good news.. or words to that effect.
>
> I don't know that its real..
>
> But the CRU code looks real

The Warring Factions

As we wrote above, there are many ways to separate the protagonists in this debate into various flavors of cowboy or Indian. In the climate debate it is inevitable that various positions will evolve and with them certain frames of mind and ways of construing the scientific issues. While there is no accepted categorization of positions, several camps or tribes have evolved and it is instructive to have some general understanding of the beliefs of these "tribes." At one end of the spectrum is a group of people who are pejoratively referred to as "alarmists." The center of this faction believes that the increase we see in global temperatures is the direct result of human activity, in particular through the addition of GHGs to the atmosphere. In addition there is a tendency for this group of individuals to call for immediate action. At times they construe the findings of the IPCC as being "conservative" and they warn that the future might even be worse than we expect. Their call is for immediate action to prevent the world from reaching a tipping point of no return. The next faction is the "warmers." "Warmers," like "alarmists" attribute the change in climate to human causes; unlike the "alarmists," they tend to make more measured comments about policy and express more uncertainty about the future course of the climate. Still, they call for action on a global basis. The next faction is referred to as the "Lukewarmers," a term coined on Climate Audit and perpetuated with the founding of Lucia Liljegren's blog, The Blackboard.

The defining characteristics of a "Lukewarmer" have emerged over time and can best be described as follows. "Lukewarmers," like "alarmists" and "warmers" believe that man's activity of adding GHGs to the atmosphere will indeed warm the planet. However, they tend to attribute the warming seen to date to a variety of sources: GHGs, land use changes, Urban Heat Island, and natural variability. With regards to policy the "Lukewarmers" take the position that actions should be taken based on the certainty of the science. Perhaps most notably, the "Lukewarmers" focus much of their effort on getting access to scientific data and methods.

Finally, the last faction is the "skeptics." That perhaps is a poor choice of names for this group of people as they have many varied beliefs and positions. However, the chief defining characteristic of a skeptic is someone who does not believe that it is getting warmer, or does not believe that increased GHGs cause the planet to warm. As alternatives they look to other causes for the changes in the temperature record.

Both of your authors consider themselves to be Lukewarmers, and we attempt to explain why at the conclusion of this book.

There is a practical consequence of segmenting beliefs in this way—alarmists in particular tend to group all opponents together, and ascribe beliefs and actions from the far extreme of skeptics to anyone who questions their positions. So although we believe that global warming is real and that action is needed to combat rising temperature, alarmists can and often do dismiss us as denialists and flat earthers. Irksome indeed.

The fear the climate scientists had was that corporations with their profit motive would use or create a corrupt publishing process to sell people doubt. To be sure, early on in the debate some of the skeptical science did have questionable funding sources. But this was not true for the skeptics that actually proved to be leaders or agenda setters—not McIntyre and not Watts, certainly.

Over time corporations have moved to get on the "climate change" bandwagon, for example by funding CRU and others, giving large sums for environmental studies at universities and donating to environmental organizations. Corporations like oil companies have the ability to transform their identity, from oil company, for example, to "energy" company. They can answer threats to their profits by changing what they do and what they sell. Scientists, on the other hand, who operate from personal motives such as pride and ego, cannot change so easily. They have their lives and identities invested in 'their' science being right. This explains why Jones can say, as he did, that he would rather we do nothing about climate change so that his science could be proven correct. It also explains why Jones would deny data to Warwick Hughes, arguing that Hughes' only aim was to find something wrong with Jones' work. Unlike an oil company which can change over time into a wind or solar power company, Jones has no such option. He is his science. The same goes for Mann. He is his hockey stick.

CHAPTER TWO: THE DATA WAS NOT IN ORDER

Cheat Sheet: To help readers through the detail we need to present to make our case, we offer these summaries in front of the remaining chapters.

The IPCC was founded to advise politicians on how bad global warming was going to be and based on the assumption that recent warming was the most extreme in history. They assume global warming is real—they do not question its existence. This leaves skeptical scientists out in the cold, for the most part.

The key to any debate about global warming is knowing with confidence what temperatures have been doing over a long enough period to make statistical projections valid. One element of Climategate is that it appears that CRU (East Anglia University's Climate Research Unit) did not do a particularly good job of taking care of the data that shows the temperature record. This has been advanced as a possible reason for Phil Jones' (director of CRU) reluctance to part with the data.

The major problems faced in creating and maintaining an accurate historical record of temperatures are essentially good housekeeping, and we see that data might be better off if placed in the hands of an objective third party, something Steve McIntyre pointed out on his weblog.

In this chapter we introduce and discuss the Urban Heat Island effect, which one of the Climategate principals, the same Phil Jones, addressed in a short but seminal paper way back in 1990. Jones thought UHI was negligible, but the data he used to justify his opinion suffered from exactly the type of good housekeeping problems that we refer to, and which plague much of the work of the Climategate scientists overall. His 1990 paper is very influential, but may be wrong, and certainly his unwillingness to cooperate with those seeking his data is unusual.

We also see that Jones was much more willing to engage with people asking for data earlier in his career. As Jones is the scientist who is now famous for saying he would delete files rather than surrender them, and who refused to send data to a scientist saying, "Why should I send these to you if you're only going to find something wrong with them?", it raises the question of what prompted the change, which coincided with his closer association with The Team, especially Michael Mann.

As the IPCC came into being there were two fundamental beliefs that structured the science that was to follow. First was the belief that the historical temperature record showed an increase in warming. The second was the belief that the warming scientists were predicting would be unprecedented in the course of human history: The 10th Anniversary brochure of the IPCC points out:

> **In 1985 a joint UNEP/WMO/ICSU Conference was convened in Villach (Austria) on the "Assessment of the Role of Carbon Dioxide and of Other Greenhouse Gases in Climate Variations and Associated Impacts". The conference concluded, that "as a result of the increasing greenhouse gases it is now believed that in the first half of the next century (21st century) a rise of global mean temperature could occur which is greater than in any man's history." It also noted that past climate data may no**

longer be a reliable guide for long term projects because of expected warming of the global climate; that climate change and sea level rises are closely linked with other major environmental issues; that some warming appears inevitable because of past activities; and that the future rate and degree of warming could be profoundly affected by policies on emissions of greenhouse gases.

The two fundamental beliefs—that greenhouse gases are warming the planet and the warming will be unprecedented—are behind the entire Climategate controversy and there are two FOIA (Freedom of Information Act) requests to investigate each of these beliefs—one targeted at the historical record of temperatures and the other aimed at understanding how Chapter 6 of the IPCC's AR4 came to be written. But for the efforts of a few scientists and engineers who sought this information, Climategate would never have happened.

Although our story's natural 'hero' (or villain, if you support The Team's position) is Steve McIntyre, our story starts back in the early 1990's with Australian researcher Warwick Hughes. Hughes was interested in two papers published by Phil Jones. One was a 1986 study on temperatures in Australia and the other a paper on what is known as the Urban Heat Island effect, or UHI, published in 1990. The UHI paper, known as *Jones 1990 et al*, is a cornerstone publication that many subsequent papers on UHI refer to and continues to be cited in IPCC reviews of science. One key paper that rests on or refers to Jones 1990, is Thomas Peterson's study in 1999. The studies form a geographic patchwork, with Jones looking at the issue in Australia, China and Russia, while Peterson was looking at the issue in the US.

More Than Mere Mortals Should Ever Be Forced To Learn About UHI

Here is one of several fairly technical sections of our book. It's important enough to include—remember that Climategate's origins involve a struggle to get information about the urban heat island effect (UHI)—and you don't need deep scientific expertise to understand what follows, but it's a bit tougher than most of the book—sorry. Let's all learn about the urban heat island effect, known as UHI.

Sidewalks Get Hotter in the Summer Sun Than Grass

The phenomenon of UHI was first identified by Luke Howard, the "father of meteorology" in 1810. Howard made detailed and comprehensive meteorological measurements in the city of London and noted that the urban area was consistently warmer than the surrounding rural areas.

The phenomenon has been the subject of detailed studies by many scientists and it is now understood that Urban Heat Islands exist because of the changes man makes to the landscape in the process of building cities. The principle cause of UHI is the building materials used in constructing cities. Concrete, asphalt, brick and other materials all retain the heat they are exposed to during daylight hours and they release that heat at night. They also change the way the air flows over the surface. This has the effect of raising the nighttime temperatures. So for example a rural area may have a daytime high temperature of 20C and a nighttime temperature of 10C which would yield an average temperature of 15C.

In contrast, an urban center in the same locale may have a daytime temperature of 20C, but at night the structures release the heat they have stored during the day and block the wind, and it may only drop to 12C, yielding an average of 16C. Thus, if a weather station were located in a rural area and then over time that rural area had become more and more urban one would see an apparent rise in temperatures from 15C to 16C. Yet, this rise would not be due to a change in the climate but rather a change in the land use.

However, there are several things that human activity can do to the land's surface to change the temperature of the air above it. It can change the albedo or the tendency of the land's cover to reflect light energy back towards the sky. A lake's surface on a quiet day has a high albedo, rough asphalt has a low albedo. It can change the emissivity or the ability of the surface to radiate the heat it has absorbed. This is quite often inversely proportional to the albedo. Getting rid of surface water or plants containing water would decrease evaporative cooling. Orienting surface structures so that they capture more sunlight would increase temperature.

Waste heat has a small, localized effect. But most of all, human activity can change the effective heat capacity of the surface, or how much energy it takes to raise the surface temperature by a given amount. Water has the highest heat capacity of any naturally occurring substance at everyday temperatures. Concrete and asphalt are much lower. Metals are lower still. Every time we clear cut a forest, drain a swamp, plow under ground cover, or pave a parking lot, we are decreasing the heat capacity of that land; so even if the same amount of sunlight hits it, it will have a higher temperature and so will the air that it contacts.

Urban heat islands should be thought of as one end of a continuum, with a climax forest or tall grass prairie at the other end. Urban heat island effect isn't something that happens overnight, it's the end of a process that starts with clear-cutting, then plowing, then paving. The idea that you can estimate the effect of UHI on the measurement of global temperature by comparing the temperature records of small towns and rural areas (as Jones did in his 1990 Nature paper) is probably not the best way of looking at this. In fact, it shows a lack of understanding of the full nature of the effect. Most everything is being developed in one way or another and most development makes the surface more like a city. Rural areas should not be thought of as controls in an experiment.

UHI can also have counter-intuitive effects on weather and climate. In a natural state the temperature of the Earth's surface changes gradually from one point to another. But once mankind starts to develop it, the factors that we mentioned above can change abruptly from one place to another. This can generate thermal convection currents much larger than what occurs naturally. The hot air that rises displaces colder air that falls. But because of the Coriolis effect, the air falling is not going in the same direction as air rising. This can cause very measurable changes in wind direction.

The UHI is a modern phenomenon and is captured to some extent by the thousands of measuring stations spread out across the world. To use this information effectively, the data has to be accurately measured and recorded, of course, but then a series of adjustments must be made to it, based on the time the temperature is recorded, whether there have been changes made to the station, what kind of instrument is measuring the temperature, etc. One of the adjustments that is made is to counter the UHI effect. One central problem in the debate on climate change is that skeptics cannot find out what stations are used in creating the temperature record, although they know it is a small subset of the whole group of stations. Another is they don't know what adjustments are made and why. Phil Jones' paper of 1990 said the adjustment for UHI could be very small. Skeptics don't agree.

To understand the issues with UHI and the climate record there is no better source than the IPCC itself. In chapter 1 of AR4 working group number 1(WG1) we find the following:

> **Inspired by the paper Suggestions on a Uniform System of Meteorological Observations (Buys-Ballot, 1872), the**
>
> **International Meteorological Organization (IMO) was formed in 1873. Its successor, the World Meteorological Organization (WMO), still works to promote and exchange standardized meteorological observations. Yet even with uniform observations, there are still four major obstacles to turning instrumental observations into accurate global time series: (1) access to the data in usable form; (2) quality control to remove or edit erroneous data points; (3) homogeneity assessments and adjustments where necessary to ensure the fidelity of the data; and (4) area-averaging in the presence of substantial gaps.**

These four problems are not issues that require much in the way of advanced or unique scientific understanding. Rather the problems belong more properly to the rather mundane tasks of data collection, quality control, and statistics. The problems faced here are the kinds of problems many engineers, statisticians and scientists in related fields face. (Another subplot to the entire Climategate story is the often expressed antipathy between engineers and academic scientists. Many very vocal skeptics are engineers, and they are perpetually puzzled at the seeming inability of 'ivory tower' academics to implement good hygiene practices in the collection, storage and dissemination of data, an issue that is key to climate controversies.)

Obstacle #1, "access to data in a usable form" is not unique to climate science; Obstacle #2, quality control is likewise not unique to climate science. Obstacle #3, homogeneity of data does require some understanding of climate measuring systems and the ways in which those measurements can be corrupted, and specifically it requires some understanding of measurement devices, thermometers, and problems such as UHI. Obstacle #4, area averaging in the presence of gaps is a purely statistical problem with a large body of standard statistical procedures one can rely on. In short, the problems of the surface record are just the kind of problems that many, including many skeptical engineers, have solved in their everyday jobs.

Posted at Climate Audit by Steve McIntyre

Posted Feb 17, 2007 at 10:44 PM | Permalink | Reply

#78. Both CRU and USHCN temperature calculations are more like accounting systems than scientific research. The science is relatively trivial compared to the accounting – and is limited to things like calculating the difference between canvas and wooden buckets.

Calculating a temperature index is a lot like calculating a Consumer Price Index. In the real world, the Consumer Price Index is calculated by a professional statistical service, not by academics writing little papers for journals. In bizarro-world, the temperature indices calculated by scientists working part-time as amateur

> accountants and obviously not doing a very good job as accountants – as witness, the lack of audit trails, the inability to locate (say) key diskettes, their either losing the original unadjusted data or their failing to take care to collect unadjusted data in the first place (or to collect/save the adjustment process where adjusted data was used.) Temperature indices should be calculated by a professional statistical service, who understand data integrity, not climate scientists who are untrained in statistical management and doing it on a seat-of-the-pants basis. Also the accounting obviously shouldn't be done by people who are also advocates. It taints the ability of third parties to trust their results.

The concern facing the scientists in the IPCC was that the historical temperature records they had compiled were showing an increase from the 1850's to the present day. Was this increase due to the changes man was making to the atmosphere or rather the simple fact that cities were being built up around the places were thermometers were placed? Specifically, is there a problem with the homogeneity of the data as noted above in "Obstacle #3?" Was the rise due to GHGs or to concrete, asphalt, and the waste heat put out by growing population, or was it due to some combination of the two? Again, from the AR4 Working Group 1 Chapter 4:

> One recurring homogeneity concern is potential urban heat island contamination in global temperature time series. This concern has been addressed in two ways. The first is by adjusting the temperature of urban stations to account for assessed urban heat island effects (e.g., Karl et al., 1988; Hansen et al., 2001). The second is by performing analyses that, like Callendar (1938), indicate that the bias induced by urban heat islands in the global temperature time series is either minor or non-existent (Jones et al., 1990; Peterson et al., 1999).

Jones et al 1990 is actually a 4 page letter published in the prestigious journal, Nature. In the paper Jones looks at sites in Western Russia, Eastern Australia, and China and concludes that urban bias in the record is very small, on the order of .05C per century, less than one tenth of the magnitude of warming seen in the instrument record. This meant for Jones' work, as well as for the many scientists who referenced his paper for their own projects, the urban heat island was insignificant as far as impacts on determining land temperatures, and has been more or less ignored since then, or at least until people like Anthony Watts and Steve McIntyre resurrected the issue.

Out of 38 stations in the Jones 1990 et al. Western Russian rural series for this region, 9 were from towns circa 10,000 population and the remaining 29 were from more remote sites. The oddity in this data is twofold. Firstly in subsequent studies a population of 10,000 is used to classify a site as peri-urban or small town, hardly a concrete jungle. So in effect Jones is comparing what would appear to be rural stations with small town stations. Nevertheless, what he reports is UHI contamination of about .12C in the period 1930-1987, about .2C per century as opposed to the .05C commonly cited as the bias due to UHI. As we shall see, the interest in data from Russia and its reliability continues to this day. And Jones' successful efforts at preventing authors from publishing results which questioned his results in Russia will also be detailed. In addition, Jones 1990 cited an Australian series of 49 stations and a Chinese network of 84. Each of these cases has its own set of irregularities, and the latter has even given rise to an investigation of academic misconduct on the part of the researcher who handled the data from China.

The Climategate files, in addition to including emails, also include documents. In this case they include what appear to be internal documents from the university (SUNY) that conducted the investigation of academic fraud lodged against Jones' co author Wang who relied on the work of researcher Zeng. Here is what they found:

> **The 84-stations are a subset of the 60- and 205-stations datasets with 35-stations from the 60-station dataset, and 49-stations from the 205-station dataset. Note that while the station history of the 35-stations was used by Professor Zeng in compiling the number of station moves in the Table included in the Appendix, the 49-stations (most rural stations) are based on her recollection (together with checking against the present-day station location), simply because the original station history manuscripts (archived at IAP) and her detailed notes were no longer available due to several office moves over the almost 19-years time span.**

The irony of this is not lost on those who were questioning Jones' study. As noted above in the IPCC AR4 Chapter 1, the number 1 and number 2 obstacles to getting a good record of the historical climate comes down to basic record keeping. Providing quality data in a usable form. And yet when Jones and his co authors set to investigate this record and seek to correct it where necessary, they proved to be prone to losing records themselves, rendering their work unaccountable.

It was the data covering eastern Australia that got Warwick Hughes' attention, according to his blog. As an Australian, Hughes had had an interest in Jones' work for some time, starting with a 1986 paper written by Jones on the problem of UHI in Australia. With his first-hand knowledge of the country, his interest in the weather, and his scientific skills, Hughes went to work trying to understand how Jones could have gotten the results he claimed. Here too is another structural element to the whole Climategate controversy. Many of those who challenge the IPCC consensus tend to have their "boots on the ground" with a deep local focus. And so much of what invigorates the conflict is the age-worn meme of the academic in the ivory tower, staying locked in that tower and not going into the field.

Hughes' interest in Jones' 1990 paper culminated in what has become a very famous email exchange. Hughes had been asking Jones for his underlying data in a number of email exchanges. The Climategate files contain an exchange between Hughes and Jones which is critical, not for its technical issues, but rather for its tone. From the Climategate files, Hughes in 2000 writes Jones:

> *At 05:13 AM 9/14/00 +1000, [warwick] wrote:*
>
> *Dear Phillip and Chris Folland (with your IPCC hat on),*
>
> *Some days ago Chris I emailed to Tom Karl and you replied re the grid cells in north Siberia with no stations, yet carrying red circle grid point anomalies in the TAR Fig 2.9 global maps. I even sent a gif file map showing the grid cells barren of stations greyed out. You said this was due to interpolation and referred me to Phillip and procedures described in a submitted paper. In the last couple of days I have put up a page detailing shortcomings in your TAR Fig 2.9 maps in the north Siberian region, everything is specified there with diagrams and numbered grid points.*
>
> *[1] One issue is that two of the interpolated grid cells have larger anomalies than the parent cells !!!!?????*
>
> *This must be explained.*

[2] Another serious issue is that obvious non-homogenous warming in Olenek and Verhojansk is being interpolated through to adjoining grid cells with no stations, like cancer.

[3] The third serious issue is that the urbanization affected trend from the Irkutsk grid cell near Lake Baikal, looks to be interpolated into its western neighbor.

I am sure there are many other cases of this, 2 and 3 happening.

Best regards,

Warwick Hughes (I have sent this to CKF)

The TAR refers to the Third Assessment Report of the IPCC, published in 2001, and Hughes is questioning some issues that he sees in the figures. Jones replies on the email thread:

From: Phil Jones <p.jones@uea.ac.uk>

To: <wsh@unite.com.au>

Subject: Re: TAR

Date: Mon Sep 18 16:23:04 2000

Cc: ckfolland@meto.gov.uk, tkarl@ncdc.noaa.gov

Warwick,

I did not think I would get a chance today to look at the web page.

I see what boxes you are referring to. The interpolation procedure cannot produce larger anomalies than neighbors (larger values in a single month). If you have found any of these I will investigate. If you are talking about larger trends then that is a different matter. Trends say in Fig 2.9 for the 1976-99 period require 16 years to have data and at least 10 months in each year. It is conceivable that at there are 24 years in this period that missing values in some boxes influence trend calculation. I would expect this to be random across the globe.

Warwick,

Been away. Just checked my program and the interpolation shouldn't produce larger anomalies than the neighboring cells. So can you send me the cells, months and year of the two cells you've found ? If I have this I can check to see what has happened and answer (1).

As for (2) and (3) we compared all stations with neighbors and these two stations did not have problems when the work was done (around 1985/6).

> *I am not around much for the next 3 weeks but will be here most of this week and will try to answer (1) if I get more details. If you have the names of stations that you've compared Olenek and Verhojansk with I would appreciate that.*
>
> *Cheers*
>
> *Phil*

In 2000 we see Jones' willingness to work with Hughes, answer his questions, double check his results, and follow up accordingly. The mails are cordial and there is no sense of animosity between the two.

By Feb 21, 2005 Jones' attitude changes radically. In July 2004 Hughes asked Jones to supply his data to support the claim that the globe had warmed .6C since the end of the 19th century. Sometime before Feb 18th, 2005 Hughes writes:

> *Dear Phil,*
>
> *Greetings from sunny Perth.*
>
> *Re the issue of the non-availability of your station by station global data. I have this month emailed the WMO a couple of times with the text below this email, no response so far. I have used the following two email addresses from the "ADMINISTRATIVE DIRECTORY Offices and Departments:" on the WMO web site. First the Office of the Secretary-General(SG) and looking down their long list of Departments I selected World Climate Data and Monitoring Programme Division (WCD) with email address . I wondered if you had any suggestion for a WMO email address that might respond ? Best wishes, Warwick Hughes*

Jones responds on the 18th:

> *Jones to Hughes, Feb 18, 2005*
>
> *Warwick,*
>
> *In Pune, India at the moment. Back in CRU on Monday but away then the rest of next week. WCDMP doesn't have a head at the moment as the last person retired in December. I will look at my email addresses I have on my machine at work on Monday. Phil*

Then Jones replied famously on Feb 21 of 2005

> *Subject: Re: WMO non respondo ... Even if WMO agrees, I will still not pass on the data. We have 25 or so years invested in the work.* **Why should I make the data available to you, when your aim is to try and find something wrong with it.** *... Cheers Phil*

Hughes had requested the underlying data that Jones had used in his reconstruction of the global temperature series. Since the WMO and its members are all committed to the open sharing of climate data and since the IPCC is a creation of the WMO, Hughes felt he was on solid ground in asking Jones for the data. However, the WMO had not responded and Jones wrote that "even if" the WMO agreed to the release of data, he would not give it to Hughes. Jones' reason for ignoring WMO guidelines is simply this: Jones thinks Hughes' **motive** is to find something wrong with it.

However, the scientific method works to produce answers the public can trust because scientists check each others' work. They work to find errors, to correct them. By sharing data and methods science moves forward through self correction. Finally, by sharing data and code, motives become irrelevant. If one's math checks out, then politics, motives, religious beliefs, sexual orientation, gender, age, race all become immaterial. Yet here Jones refuses to comply with the very fundamentals of science. Instead of releasing data which removes the specter of hidden motive, Jones turns to questioning the motives of others. **What changed the tone between 2000 and Feb 21, 2005? What altered Jones' willingness to exchange information with other people? The publication of one paper in 2003 and the bitter rivalry it engendered.**

CHAPTER THREE: THE DEVIL AND MR. JONES

Cheat Sheet: The cutoff date for the big IPCC report that will be published in 2007 is actually in 2005. Papers have to be published in peer-reviewed journals by then in order to be featured in the IPCC's AR4, a hugely important report. Because McIntyre has successfully challenged Michael Mann's Hockey Stick chart (which was the poster child of the previous IPCC report and spread instantly across the world), The Team needs to find a replacement source of data in time to be included in the report. Recall that from a political perspective it is important to show that the recent rise in temperatures is unprecedented, which means that the Medieval Warming Period (which for generations was thought of as warmer than today) is more than inconvenient. The Hockey Stick made the MWP disappear—but because of McIntyre's challenges, the Hockey Stick is 'broken.' It is also important politically to preserve the sanctity of the modern temperature record as collected by land stations—it is this data, after all, that shows the modern temperature rise to best effect. Questions about how temperatures are adjusted and the correct levels for the urban heat island effect are not at all welcome.

The scientists who form The Team, and some of their colleagues, are under pressure to find a reconstruction of ancient climate to support or replace the Hockey Stick, get it published and peer-reviewed, while fending off more and stronger criticism in the media (and from McIntyre). The emails hint at some of the effects—getting more defensive, dismissing critics as 'frauds,' seeing the entire world in an 'us vs. them' frame of mind, and taking active steps against those they perceive as enemies.

Two of The Team publish a paper that attempts to counter McIntyre's criticism of Mann, but the race is on to get it peer-reviewed and published in time to enter the IPCC report. Fortunately, they have friends on the inside. In fact, the editor of the IPCC chapter concerned happens to be a member of The Team.

Chapter 3 details the change in attitude on Jones' part about sharing data. His attitude in 2002 is changed by the publication of McIntyre's 2003 paper and Mann's diatribes against McIntyre. In 2004, with McIntyre's work and criticism gaining traction on the internet, the climate scientists launch their PR counter offensive: the weblog Real Climate. As 2004 comes to a close, McIntyre and McKitrick get a paper accepted into the prestigious journal GRL, which leads Mann to suggest a campaign against the editor that really seems right out of the playbook of Joe McCarthy. Jones, Wigley and Mann all discuss the danger that FOIA poses, and before the first FOIA is even issued Jones threatens to delete data if McIntyre ever finds out about FOIA. The McIntyre paper gets published mid Feb 2005. Mann is getting increasing notoriety for secreting data away, and Jones decides to follow Mann's lead and denies a request for his data, data he previously thought should be public.

In 2003 Stephan McIntyre and Ross McKitrick published "Corrections to the Mann et al (1998) Proxy Data Base and Northern Hemisphere Average Temperature Series" *Energy and Environment* 14(6) 751-772.

This paper (MM03) purported to critique an influential paper (Mann, Bradley and Hughes, 1998) and correct an influential scientist (Michael Mann) on the issue of climate reconstruction. Mann's work served to flatten the temperature record of the past 1,000 years, almost eliminating both the

Medieval Warming Period and the Little Ice Age. In particular, this buttressed the IPCC claims that the current warming is greater than any seen in the history of man. McIntyre's critical paper created an animosity between Mann and McIntyre that later shaped the profile of the Climategate story and Phil Jones, who hitherto had worked willingly with Hughes and McIntyre, was transformed from someone who would share data, even when it might be covered by confidentiality agreements, to someone who would rather delete it than share it.

Jones was there when Mann first learned of the paper that was to become known as MM03, and very quickly he and his colleagues in the ivory tower were called to do battle in a media and political war.

In October of 2003 Mann was passed a document that leaked word of the MM03 paper.

> *Two people have a forthcoming 'Energy & Environment' paper that's being unveiled tomorrow (Monday) that -- in the words of one Cato / Marshall/ CEI type -- "will claim that Mann arbitrarily ignored paleo data within his own record and substituted other data for missing values that dramatically affected his results. When his exact analysis is rerun with all the data and with no data substitutions, two very large warming spikes will appear that are greater than the 20th century. Personally, I'd offer that this was known by most people who understand Mann's methodology: it can be quite sensitive to the input data in the early centuries. Anyway, there's going to be a lot of noise on this one, and knowing Mann's very thin skin I am afraid he will react strongly, unless he has learned (as I hope he has) from the past...."*

Remember that the central thrust of Mann's work is to show that recent temperature rises are dramatic and unprecedented, and much of this is accomplished by showing that the Medieval Warming Period and Little Ice Age were less dramatic than previously thought—doing this gave a flattened shape to the 'shaft' of Mann's Hockey Stick, and made the rise in the 'blade' appear even more dramatic.

Whoever passed this document on to Mann was giving him good advice, especially with regards to his temperament, but Mann would not heed this counsel. Mann proceeded to line up his forces for battle, sending off the following mail titled CONFIDENTIAL with Phil Jones as a recipient. Mann cc's himself on the mail as well and copies the above message from the mystery third party. This parenthetically suggests that McIntyre's request seemed relevant to all, and that the issue of providing data would be of concern to respected scientists from a variety of prestigious institutions studying climate change—or at least that Mann felt they would share his concerns.

> *From: "Michael E. Mann" <mann@virginia.edu>*
>
> *Subject: CONFIDENTIAL Fwd:*
>
> *Date: Sun, 26 Oct 2003 13:47:44 -0500*
>
> *Dear All,*
>
> > *This has been passed along to me by someone whose identity will remain in confidence. Who knows what trickery has been pulled or selective use of data made. It's clear that "Energy and Environment" is being run by the baddies--only a shill for industry would*

have republished the original Soon and Baliunas paper as submitted to "Climate Research" without even editing it. Now apparently they're at it again...

My suggested response is: 1) to dismiss this as stunt, appearing in a so-called "journal" which is already known to have defied standard practices of peer-review. It is clear, for example, that nobody we know has been asked to "review" this so-called paper 2) to point out the claim is nonsense since the same basic result has been obtained by numerous other researchers, using different data, elementary compositing techniques, etc. Who knows what sleight of hand the authors of this thing have pulled. Of course, the usual suspects are going to try to peddle this crap. The important thing is to deny that this has any intellectual credibility whatsoever and, if contacted by any media, to dismiss this for the stunt that it is..

Thanks for your help,

mike

Mann starts his mail with a suspicion of trickery, rather than a presumption that perhaps he or the person who questioned his results has made an innocent mistake. And he follows with a quick appeal to ulterior motives. Simply, if you contest his views you must be funded by industry. Then he notes that it's obvious that no one "we know" has been asked to review it. Next he moves to the media strategy for the team and gives them the party line. Sadly, for Mann his media strategy skills were lacking. It would turn out that the authors of the paper, while not perfect, were no amateurs and were quite able to show that their paper was not a stunt.

Four days later Ray Bradley, one of Mann's co authors, intervenes and suggests another strategy. It's notable because it rests on subterfuge. Essentially at a time when Climate Science is coming into question, when trust is a key issue, Bradley suggests the following dodge:

Date: Thu, 30 Oct 2003 11:55:18 -0500

Cc: mann@multiproxy.evsc.virginia.edu, mhughes@ltrr.arizona.edu

Tim, Phil, Keef:

I suggest a way out of this mess. Because of the complexity of the arguments involved, to an uniformed observer it all might be viewed as just scientific nit-picking by "for" and "against" global warming proponents. However, if an "independent group" such as you guys at CRU could make a statement as to whether the M&M effort is truly an "audit", and if they did it right, I think that would go a long way to defusing the issue. ....

If you are willing, a quick and forceful statement from The Distinguished CRU Boys would help quash further arguments, although here, at least, it is already quite out of control.....yesterday in the US Senate the debate opened on the McCain-Lieberman bill to control CO2 emissions from power plants. Sen Inhofe stood up & showed the M & M figure and stated that Mann et al--& the IPCC assessment --was now disproven and so there was no reason to control CO2 emissions.....I wonder how many times a "scientific" paper gets reported on in the Senate 3 days after it is published....

Ray

Not coincidentally, the address list is much shorter, suggesting that only certain people could be trusted with this kind of subterfuge. .

Ray Bradley is asking the scientists at CRU to pretend they are an "independent group" of distinguished scientists and hoping the stunt would fool reviewers. This too would fail as a media strategy and before it was over, respected statistician Edward Wegman would demonstrate the rather incestuous relationships between Mann and the rest of The Team.

Further evidence of the change in Jones' attitude towards sharing data can be seen in the examination of his correspondence with McIntyre prior to the publication of that 2003 paper. What follows is correspondence, toward the end of 2002, between Jones and McIntyre prior to McIntyre's publication in 2003

sept 8 2002 From Stephan McIntyre forwarded to P Jones:

In Journal of Climate 7 (1994), Prof. Jones references 1088 new stations added to the 1873 stations referred to in Jones 1986. Can you refer me to a listing of these stations and an FTP reference to the underlying data?

Thanks, Steve McIntyre

Dear Steve, You are looking into station lists from papers in the early 1990s and 1980s. These are now out of date. There will be a new paper coming out in J. Climate (probably early next year). I'm attaching the station list (5159 stations) from that paper ……… Once the paper comes out in the Journal of Climate, I will be putting the station temperature and all the gridded databases onto our web site. The gridded files on our web site at the moment are from our current analysis. The new analysis doesn't change the overall character of the gridded fields, it is just easier for me to send the new lists of stations used from the new analysis. I hope this helps.

Phil Jones

Dear Steve, Attached are the two similar files [normup6190, cruwld.dat] to those I sent before which should be for the 1994 version. This is still the current version until the paper appears for the new one. As before the stations with normal values do not get used.

I'll bear your comments in mind when possibly releasing the station data for the new version (comments wrt annual temperatures as well as the monthly). One problem with this is then deciding how many months are needed to constitute an annual average. With monthly data I can use even one value for a station in a year (for the month concerned), but for annual data I would have to decide on something like 8-11 months being needed for an annual average. With fewer than 12 I then have to decide what to insert for missing data. Problem also applies to the grid box dataset but is slightly less of an issue.

I say possibly releasing above, as I don't want to run into the issues that GHCN have come across with some European countries objecting to data being freely available. I would like to see more countries make their data freely available (and although these monthly averages should be according to GCOS rules for GAA-operational Met. Service.

Cheers Phil Jones

Like Hughes, McIntyre was interested in the problem of UHI. In fact, his entry into the field of climate science started with the issue of UHI rather than the issue of climate reconstruction. In 2002 while the Kyoto treaty was being discussed, McIntyre had a lunch discussion with a friend who was a geologist. The friend mentioned that he thought the time scale the climate scientists were considering was very short in geological time and that the warming was nothing to take note of. McIntyre, motivated by nothing more than curiosity, started reading the IPPC's TAR (Third Assessment Report, and that led him to Jones' 1990 paper. Were the temperature records infected by a spurious warming caused by UHI?

In order to understand this, McIntyre asked Jones for the underlying data. Jones was cooperative, open to suggestions, and helpful in providing information. In fact, Jones released a **version** of some of the data at the heart of the Climategate controversy, all the while noting that the data might be covered by confidentiality agreements, and he notes per WMO guidelines that this data should be publically available from members of the WMO. So as late as 2002, Jones was willing to share his data, even though he believed it might be covered by agreements and he held the position that this data should be released, citing WMO guidelines. Yet as noted above on Feb 21, 2005, Jones changed his mind about WMO guidelines when he refused to release data to Hughes, even if WMO agreed to release it.

MM03 attacked one of the central papers that supported a tenet of climate science, that today's temperatures were unprecedented. The battle between Mann and McIntyre was heated, but up until Oct 14th, 2004 Mann and the other climate scientists held the upper hand, as they could argue that McIntyre had not published in a mainstream journal, as MM03 had been published in Energy and Environment, a journal not respected in the climate science community. However, on October 14th the journal Geophysical Research Letters, (GRL) received the following paper from McIntyre and McKitrick:

Hockey sticks, principal components, and spurious significance, Stephen McIntyre Northwest Exploration Co., Ltd., Toronto, Ontario, Canada, Ross McKitrick Department of Economics, University of Guelph, Guelph, Ontario, Canada. Received 14 October 2004; revised 22 December 2004; accepted 17 January 2005; published 12 February 2005.

Note the date. The paper was received, Oct 14th 2004. It was under review during the late fall of 2004 and published eventually on Feb 12th 2005. During this time frame Jones and other climate scientists, most notably Mann, were becoming increasing concerned with some of the messages being delivered in the press, especially when those stories were covering climate reconstructions. On Oct 14th, 2004 Jones writes to Mann about another critic, Hans von Storch and an email he has received on the 13th from a Dutch journalist who cites an online version of a paper.

At 15:28 13/10/2004,

Dear professor Jones,

(We met ten days ago in Utrecht, when Albert Klein Tank got his PhD). I am a science journalist of the Dutch daily newspaper NRC Handelsblad in Rotterdam ([1]www.nrc.nl). I try to write an article about climate (surface temperature) reconstruction as far back as the year 1000 - the well know Mann, Bradley, Hughes (1998 and 1999) research. The reason is, of course, the publication of the article of Von Storch, Zorita, c.s. in Science-online (30 september). Von Storch claims that the statistical approach of Mann c.s. produced a serious underestimation of the low

frequency (long term) oscillations in global temperature. The conclusion could be that the Medieval Warm Period was in fact warmer than today. And the recent warming is - after all - not so special. Can you in a few words - and for a general public - give a comment on the paper? Does it make sense? It seems pretty convincing to me. Can you help me?

Waiting for your reply,

sincerely yours,

Karel Knip

NRC Handelsblad

Jones writes to Mann on the 14th complaining about the way von Storch is handling journalists and we see what is the beginning of scientists starting to be concerned about how the media, particularly the online media, is handling the science.

From: Phil Jones <p.jones@uea.ac.uk>

To: mann@virginia.edu

Mike,

FYI. I met this guy in Utrecht last week at Albert Klein Tank's PhD ceremony. It appears from many media reports that people really believe that their run is an ALTERNATE to yours – based on no proxy data. Even Hans has sent an email around to this effect, but he obviously isn't making it as clear as I've just done to this Dutch journalist. I think he might be being clear with fellow scientists and economical with the truth with journalists, i.e. not directing them down the correct path when he sees them going down the wrong one. I should see Ray next week in Seattle at a DoE meeting.

Cheers, Phil

In October of 2004 McIntyre and his criticism was on the radar of climate scientists. Tom Wigley writes Phil Jones about McIntyre's and McKitrick's work (MM03) which is making its way around the internet. Wigley is not as dismissive of McIntyre's and McKitrick's work as is Michael Mann. In fact, Wigley calls Mann's paper a very sloppy piece of work…

At 20:46 21/10/2004, [Tom Wigley]

Phil,

I have just read the M&M stuff crtcizing MBH. A lot of it seems valid to me. At the very least MBH is a very sloppy piece of work -- an opinion I have held for some time. Presumably what you have done with Keith is better? -- or is it? I get asked about this a lot. Can you give me a brief heads up? Mike is too deep into this to be helpful.

Tom.

As Wigley notes M & M (McIntyre and McKitrick) have some valid points in their criticism of MBH (Mann and his co authors 1998 paper). What Mann viewed as a stunt others found merit in. Wigley asks Jones about his reconstruction work with colleague Keith Briffa. Briffa, as the Climategate mails and his studies show, was less certain about reconstructions of the MWP than Mann was. Jones, of course, is stuck between supporting Briffa or Mann, both co-authors. Most importantly Wigley recognizes that Mann is too deep in this to be helpful. Mann has too much at stake to be objective. Jones replies, by this time taking on some of Mann's attitudes toward McIntyre and McKitrick:

> *From: Phil Jones p.jones@uea.ac.uk*
>
> *To: Tom Wigley wigley@cgd.ucar.edu*
>
> *Tom,*
>
> *The attached is a complete distortion of the facts. M&M are completely wrong in virtually everything they say or do. I have sent them countless data series that were used in the Jones/Mann Reviews of Geophysics papers. I got scant thanks from them for doing this - only an email saying I had some of the data series wrong, associated with the wrong year/decade. I wasted a few hours checking what I'd done and got no thanks for pointing their mistake out to them. If you think M&M are correct and believable then go to this web site*
>
> *Point I'm trying to make is you cannot trust anything that M&M write. ....*
>
> *Bottom line - there is no way the MWP (whenever it was) was as warm globally as the last 20 years. There is also no way a whole decade in the LIA period was more than 1 deg C on a global basis cooler than the 1961-90 mean. This is all gut feeling, no science, but years of experience of dealing with global scales and variability.*
>
> *Cheers*
>
> *Phil*

Jones' "gut feeling" is at stake and he is clearly agitated by his encounters with McIntyre, a marked difference from their exchange in 2002. In 2002, McIntyre was merely a researcher asking for data, but by 2003 McIntyre was a published author leveling criticisms at Jones' co author Michael Mann. Jones also refers Wigley to a web site that discussed M&M. The fight over MM03 was largely taking place on the web as McIntyre had started to write about his findings on a blog called www.climate2003.com. For independent researchers like McIntyre, posting articles on the internet was far more expedient than publishing in page limited journals. And just as citizen-journalists had transformed print journalism with the advent of blogs, climate science looked ripe to be transformed by the internet. McIntyre and McKitrick also adopted a publication model used by econometricians: they posted their data and their code so that anyone could check their work, find errors and suggest improvements. This gave them the moral high ground of transparency as opposed to Mann's and Bradley's shadowy world of "independent scientists," although Mann and Bradley would certainly argue with some legitimacy that they were only following a century-old practice.

By December of 2004 the MM05 paper is under review in a journal that the climate scientists respect: GRL. As noted above, the new paper was received on Oct 14th and was under review. The importance of having a paper in what is known as "peer reviewed literature" is crucial to understand. The IPCC assessment reports are not scientific documents. They are literature reviews, that is, reviews of papers that are published in peer reviewed journals such as GRL. The fall of 2004 and early spring of 2005 is a critically important period in the preparation of the other nexus of the Climategate files: The IPCC Assessment Report 4 (AR4) Working Group 1 (WG1) Chapter 6, (Ch06), the chapter on climate reconstructions. As MM05 comes into publication, the climate scientists are drafting the very chapter that MM05 will critique. Michael Mann writes Keith Briffa, who is the lead author of WG1 AR4 Ch06, and refers to a paper authored by "Casper Ammann and Eugene Wahl that can be used in AR4 Ch06 to counter the claims made by M&M. This paper and the cases surrounding it forms the second thread of FOIA requests initiated by McIntyre.

From: "Michael E. Mann" <mann@virginia.edu>

To: Keith Briffa <k.briffa@uea.ac.uk>

Subject: email #2: paper in review in J. Climate (as a letter), discrediting McIntyre and McKitrick

Date: Mon, 13 Dec 2004 11:47:26 -0500

Keith,

This paper [Ammann & Wahl] is in review, and can be referred to (just clear w/ Caspar or Gene first) for IPCC draft purposes. They basically show that the McIntyre and McKitrick paper is total crap, and they provide an online version of the Mann et al method (and the proxy data), so individuals can confirm for themselves...

Mike

So, prior to the Ammann paper being fully reviewed accepted and published, Mann is poisoning the well with lead author Briffa. It is also important to note that the upcoming Ammann paper will post its data and methods on line, in the style of McIntyre and McKitrick, as Mann and others are slowly coming to the understanding that the internet not only allows this but creates considerable momentum driving it. People want to check for themselves.

As the year closes out Michael Mann writes Jones, calls McIntyre a fraud and counsels him on how to deal with McIntyre. He attaches a copy of the paper. It's 2004. Three years later this paper will finally be published, but for now it serves its purpose. Briffa, Jones's co worker and co-author has the paper. Jones now has the paper and both are prejudiced by Mann's opinion. If Jones was, as the record shows, willing to share data with McIntyre in 2002 by the end of 2004 it is clear that Mann is doing everything he can to cut off those communications. Mann raises the false issue of McIntyre's funding. The record shows that McIntyre receives no funding from any sources, yet Mann repeats a mantra heard on the internet and possibly fed to him by public relations consultants.

Date: Thu, 30 Dec 2004 09:22:02 -0500

To: Phil Jones <p.jones@uea.ac.uk>

> From: "Michael E. Mann" <mann@virginia.edu>
>
> Phil,
>
> I would immediately delete anything you receive from this fraud. [McIntyre] You've probably seen now the paper by Wahl and Ammann which independently exposes McIntyre and McKitrick for what it is--pure crap. Of course, we've already done this on "RealClimate", but Wahl and Ammann is peer-reviewed and independent of us. I've attached it in case you haven't seen (please don't pass it along to others yet). It should be in press shortly. Meanwhile, I would NOT RESPOND to this guy. As you know, only bad things can come of that. The last thing this guy cares about is honest debate--he is funded by the same people as Singer, Michaels, etc... Other than this distraction, I hope you're enjoying the holidays too...
>
> talk to you soon,
>
> mike

But McIntyre's criticism isn't 'crap,' and the walls created in this time frame serve to delay resolution of this issue and contribute to Climategate's impact on the debate, especially at Copenhagen in December 2009. Had it been resolved earlier, who knows what would have happened?

2005 FOIA: Jones's nightmare

As 2004 turns into 2005 the work on Chapter 6 of AR4 continues and Jonathan Overpeck gives Briffa guidance on his page count and passes along some concerns from Tim Rind. With few pages to work with Briffa would be under considerable pressure to trim anything that isn't deemed essential, but the oddity here is that the scientific certainty around climate reconstructions is low, and it would seem should engender longer, more detailed chapters rather than a short one.

> From: Jonathan Overpeck <jto@u.arizona.edu>
>
> To: k.briffa@uea.ac.uk
>
> Subject: Fwd: Re: [Wg1-ar4-ch06] IPCC last 2000 years data
>
> Date: Tue, 4 Jan 2005 21:52:47 -0700
>
> Hi Keith ....
>
> as leader of this KEY section [AR4 WG1 ch06], we need you to take the lead integrating everything you think should be integrated, editing and boiling it down to just ca 4 pages of final text (e.g., 8 pages of typed text plus figs). .... 3) Your section is too long and needs to be condensed. Thus, you need to think through what's most important and what's less so. ....

At one point in the email exchanges between Briffa and Overpeck, Overpeck would write that they don't have room to show confidence intervals. Overpeck underlines concerns from David Rind to Briffa that center on several key points, but primarily on the uncertainties and how the document presents them. Rind counsels a forthright approach rather than squirreling away commentary in a footnote or reference. And Rind notes the internet phenomenon and McIntyre. In the end Rind's recommendations to be forthright would be ignored.

> ******* *From David Rind 1/4/05* ***************
>
> *pp. 8-18: The biggest problem with what appears here is in the handling of the greater variability found in some reconstructions, and the whole discussion of the 'hockey stick'.*
>
> *The tone is defensive, and worse, it both minimizes and avoids the problems. We should clearly say (e.g., page 12 middle paragraph) that there are substantial uncertainties that remain concerning the degree of variability - warming prior to 12K BP,[12,000 years before present] and cooling during the LIA,[Little Ice Age] due primarily to the use of paleo-indicators of uncertain applicability, and the lack of global (especially tropical) data. Attempting to avoid such statements will just cause more problems.*
>
> *….. The discussion of uncertainties in tree ring reconstructions should be direct, not referred to other references - it's important for this document. How the long-term growth is factored in/out should be mentioned as a prime problem. …..*
>
> *The primary criticism of McIntyre and McKitrick, which has gotten a lot of play on the Internet, is that Mann et al. transformed each tree ring prior to calculating PCs [principle Components] by subtracting the 1902-1980 mean, rather than using the length of the full time series (e.g.,1400-1980), as is generally done. M&M claim that when they used that procedure with a red noise spectrum, it always resulted in a 'hockey stick'. Is this true? If so, it constitutes a devastating criticism of the approach; if not, it should be refuted. While IPCC cannot be expected to respond to every criticism a priori, this one has gotten such publicity it would be foolhardy to avoid it.*
>
> *especially not in an off-handed way.)*
>
> ******* *END From David Rind 1/4/05* ***************

Rind recognizes the power of McIntyre's criticism. First, Mann had misused a statistical technique (subtracting the 1902-1980 mean) and secondly, Mann's results were merely a methodological artifact. Like Wigley, Rind sees the issues with Mann's work.

Finally on Jan 21 Mann, Malcolm Hughes and Tom Wigley discuss the McIntyre paper that is coming out in GRL. (MM05) They discuss the best way to respond and bemoan the fact that they cannot control what comes out in GRL: Mann, always on the hunt for conspiracies and political solutions to scientific problems, suggests in an email eerily reminiscent of Richard Nixon, that they build a file on the editor of GRL:

> *just a heads up. Apparently, the contrarians now have an "in" with GRL. This guy Saiers has a prior connection w/ the >> > University of Virginia Dept. of Environmental Sciences that causes me >> > some unease.*

> *"I'm not sure that GRL can be seen as an honest broker in these debates anymore, and it is probably best to do an end run around GRL now where possible. They have published far too many deeply flawed contrarian papers in the past year or so. There is no > possible excuse for them publishing all 3 Douglass papers and the Soon et al paper. These were all pure crap.*
>
> *... basically this is just a heads up to people that something might be up here. What a shame that would be. It's one thing to lose "Climate Research". We can't afford to lose GRL. I think it would be useful if people begin to record their experiences w/ both Saiers and potentially Mackwell (I don't know him--he would seem to be complicit w what is going on here). If there is a clear body of evidence that something is amiss, it could be taken through the proper channels. I don't that the entire AGU hierarchy has yet been compromised!*
>
> (Wigley adds) *"Proving bad behavior here is very difficult. If you think that Saiers is in the greenhouse skeptics' camp, then, if we can find documentary evidence of this, we could go through official AGU channels to get him ousted. Even this would be difficult.*

As we will see later Mann and company are not so concerned about having honest brokers at journals. They feel they to manipulate the peer review system to get certain papers into publication. Papers they need in time for the IPCC AR4.

Although this does seem like petty back-biting between academics, the IPCC is using the materials debated here to create summaries for politicians and policy makers that will influence national energy policies the world over and lead to proposals for substantial spending to combat climate change. The stakes are much higher than individual careers and reputations, although these also form part of the controversy.

Coming to Grips With Freedom of Information

Both the United States and the United Kingdom have legislation about the freedom of information which requires public bodies to provide information to citizens when requested. There are exceptions, of course, and limitations, but the intent of the laws are clear—citizens have a right to the information used to make decisions affecting their lives.

Feb 21, 2005 comes quick on the heels of the following late January email exchange where the issue of FOIA is raised: An FOIA request is a request to a public institution that information held by that institution be released to the public. Herein follows a discussion in January of 2005 between Jones and Tom Wigley. Wigley has been sent a brochure on the FOIA act from Jones' employer, CRU at the University of East Anglia. At the time of this email, no FOIA request has even been written, yet the scientists are concerned about it and plan ridiculous strategies to avoid it.

At 02:59 21/01/2005, [wigley]

Phil,

Tom Karl told me you will be on the VTT review panel. ..... I got a brochure on the FOI Act from UEA. [University of East Anglia, Jones' university] Does this mean that, if someone asks for a computer program we have to give it out?? Can you check this for me (and Sarah). I will be at CRU next Mon, Tue, Wed in case Sarah did not tell you.

Thanks, Tom.

Phil Jones replies and appears to indicate that UEA and CRU employees do not understand the FOIA, even though the University has published guidelines and holds training courses on FOIA. Had Jones attended the training it would have been clear to him that data and code and his emails would be subject to FOIA. In addition, he would have been instructed to insure that his emails were professional:

Tom, ... On the FOI Act there is a little leaflet we have all been sent. It doesn't really clarify what we might have to do re programs or data. Like all things in Britain we will only find out when the first person or organization asks. I wouldn't tell anybody about the FOI Act in Britain. I don't think UEA really knows what's involved. As you're no longer an employee I would use this argument if anything comes along. I think it is supposed to mainly apply to issues of personal information - references for jobs ....

Cheers Phil

Jones is acting beyond the scope of his position and offering FOI advice to Wigley, a former employee. That fact that Wigley is no longer employed by UEA has no bearing on whether or not his documents fall under its guidelines. Jones certainly seems to be looking for ways to hide from complying with the law rather than complying with it. Wigley is no better in his response, either in his understanding of the law or his willingness to try to find ways to not comply with it, arguing in ridiculous fashion that an employee could "claim" that she had only written 1/10 of the code, thereby allowing the organization to release 1 out of every 10 lines.

Tom Wigley

Phil, Thanks for the quick reply. The leaflet appeared so general, but it was prepared by UEA so they may have simplified things. From their wording, computer code would be covered by the FOIA. My concern was if Sarah is/was still employed by UEA. I guess she could claim that she had only written one tenth of the code and release every tenth line. ....

. Best wishes, Tom.

Jones replies, trying to allay the fears of Wigley, who is very concerned about the release of the code. In general, as the code released in the Climategate files indicates, Wigley had good reason to be concerned, as the programmer comments and bugs found within days of release on the Internet show, indicating confusion, incompetence and a willingness to invent data.

> *From: Phil Jones <p.jones@xxxxxxxx.xxx> To: Tom Wigley <wigley@xxxxxxxx.xxx> Subject: Re: FOIA Date: Fri Jan 21 15:20:06 2005 Cc: Ben Santer <santer1@xxxxxxxx.xxx>*
>
> *Tom, I'll look at what you've said over the weekend re CCSP. I don't know the other panel members. I've not heard any more about it since agreeing a week ago. As for FOIA Sarah isn't technically employed by UEA and she will likely be paid by Manchester Metropolitan University. I wouldn't worry about the code. If FOIA does ever get used by anyone, there is also IPR to consider as well. Data is covered by all the agreements we sign with people, so I will be hiding behind them. I'll be passing any requests onto the person at UEA who has been given a post to deal with them.*
>
> *Cheers Phil*

Jones' attitude toward releasing data has changed entirely from his 2002 email exchanges with McIntyre. In those exchanges he considered the agreements to be trumped by the dictates of the WMO and he did promise to send a version of the data to McIntyre. Now he considers the agreements to be a device to hide behind, even though FOIA allows CRU to release confidential data if a public interest test is met. He also wants to hide behind the intellectual property rights (IPR) with respect to the code. In Jones' world all the software is created at taxpayer expense. In fact, it was US taxpayers, via the Department of Energy (DOE) who paid for the Jones's work. If the DOE would allow Jones to assert IPR with respect to work they paid for, then in effect they would be funding one scientist to have a monopoly on methods he had created on their dime.

As February 2005 starts Mann writes Jones and informs him of his move to Penn State, Jones replies and introduces the specter of FOIA to Mann.

> *At 09:41 AM 2/2/2005, Phil Jones wrote:*
>
> *Mike,*
>
> *I presume congratulations are in order - so congrats etc !*
>
> *Just sent loads of station data to Scott. Make sure he documents everything better this time ! And don't leave stuff lying around on ftp sites - you never know who is trawling them. The two MMs [McIntyre and Mckitrick] have been after the CRU station data for years. If they ever hear there is a Freedom of Information Act now in the UK, I think I'll delete the file rather than send to anyone. Does your similar act in the US force you to respond to enquiries within 20 days? - our does ! The UK works on precedents, so the first request will test it. We also have a data protection act, which I will hide behind. Tom Wigley has sent me a worried email when he heard about it - thought people could ask him for his model code. He has retired officially from UEA so he can hide behind*

that. IPR should be relevant here, but I can see me getting into an argument with someone at UEA who'll say we must adhere to it !

Jones is referring to an incident where McIntyre had discovered some of the data that Mann had refused to release to him. Scott Rutherford, Mann's associate, had inadvertently left the data and some code in the open on an FTP site. Also, Jones recounts McIntyre's 2002 request for Jones 1990 data, misremembering that Mckitrick had nothing to do with this request. Note finally Jones' attitude toward FOIA legislation and his worries that he will be forced to adhere to it. Later, in 2008, Jones's warning to Mann about leaving data on FTP sites will prove to be cruelly ironic, as Jones will leave an early version of his dataset in plain sight where McIntyre will find it, much to the delight of his readers.

Jones continues

Are you planning a complete reworking of your paleo series? Like to be involved if you are. Had a quick look at Ch 6 on paleo of AR4. The MWP [Medieval Warm Period] side bar references Briffa, Bradley Mann, Jones, Crowley, Hughes, Diaz - oh and Lamb ! Looks OK, but I can't see it getting past all the stages in its present form. MM and SB get dismissed. All the right emphasis is there, but the wording on occasions will be crucial. I expect this to be the main contentious issue in AR4. ....

Cheers

Phil

As Jones notes, he was looking at a draft of Chapter 6 of IPCC's AR4. And before the chapter has even been reviewed he has determined that McIntyre's work will be dismissed, but more importantly he recognizes its importance in the 4th Assessment. The exchange ends with Mann discussing FOIA in the US.

Thanks Phil,

Yes, we've learned our lesson about FTP. We're going to be very careful in the future what gets put there. Scott really screwed up big time when he established that directory so that Tim could access the data. Yeah, there is a freedom of information act in the U.S., and the contrarians are going to try to use it for all its worth. But there are also intellectual property rights issues, so it isn't clear how these sorts of things will play out ultimately in the U.S. I saw the paleo draft (actually I saw an early version, and sent Keith some minor comments). It looks very good at present--will be interesting to see how they deal w/ the contrarian criticisms--there will be many. I'm hoping they'll stand firm (I believe they will--I think the chapter has the right sort of personalities for that)...Will keep you updated on stuff...

talk to you later,

mike

By this time word of the new paper by McIntyre, set for publication on Feb 12th has reached the NYT. And environmental journalist Andrew Revkin seeks Mann's opinion and offers to send Mann the copy of a Nature paper that is not yet released

> At 02:14 PM 2/4/2005, Andy Revkin wrote:
>
> Hi all, There is a fascinating paper coming in Nature next week (Moberg of Stockholm Univ., et   al) that uses mix of sediment and tree ring data to get a new view of last 2,000 years.   Very warped hockeystick shaft (centuries-scale variability very large) but still   pronounced 'unusual' 1990's blade.   I'd like your reaction/thoughts for story I'll write for next Thursday's Times. also, is there anything about the GRL paper forthcoming from Mc & Mc that warrants a   response?   I can send you the Nature paper as pdf if you agree not to redistribute it (you know the embargo rules).   that ok?
>
> thanks for getting in touch!   Andy

Mann answers Revkin and seems unable to moderate his response. Mann accuses McIntyre of fraud, even though the entire dataset and methods for MM05 are available to all readers. Ironically Mann does not realize that by publishing their data and code McIntyre and McKitrick gain the moral high ground in the debate, for it is only when the data and methods are hidden that concerns about fraud have any weight.

We should tell readers new to this long-running story that McIntyre and McKitrick were judged to be correct in their criticism by a panel tasked with examining the controversy by a U.S. Congressional committee. The panel, led by Edward Wegman, a respected statistician, not only agreed with M&M's criticisms, but predicted that the small nature of The Team and their incestuous inter-relationships would lead to something that is very much like what we describe in this book.

> Date: Fri, 04 Feb 2005 15:52:53 -0500
>
> To: Andy Revkin <anrevk@nytimes.com>
>
> From: "Michael E. Mann" <mann@virginia.edu>
>
> Subject: Re: FW: "hockey stock" methodology misleading
>
> Hi Andy,   The McIntyre and McKitrick paper is pure scientific fraud. I think you'll find this reinforced by just about any legitimate scientist in our field you discuss this with.

Mann's strategy of screaming fraud when his work is challenged does not gain much traction in the press, particularly since Mann refuses to release any of his data or methods. On Feb 21, an apparently concerned Keith Briffa, worried perhaps that he will have to deal with MM05 in writing chapter 6 of AR4 sends a list of news clippings to Jones.

Date: Mon, 21 Feb 2005 15:40:05 +0000

To: p.jones@uea.ac.uk

From: Keith Briffa <k.briffa@uea.ac.uk>

Subject: Fwd: CCNet: PRESSURE GROWING ON CONTROVERSIAL RESEARCHER TO DISCLOSE SECRET DATA

This should have produced a healthy scientific debate. Instead, Mr. Mann tried to shut down debate by refusing to disclose the mathematical algorithm by which he arrived at his conclusions. All the same, Mr. Mann was forced to publish a retraction of some of his initial data, and doubts about his statistical methods have since grown. --The Wall Street Journal, 18 February 2005

But maybe we are in that much trouble. The WSJ highlights what Regaldo and McIntyre says is Mann's resistance or outright refusal to provide to inquiring minds his data, all details of his statistical analysis, and his code. So this is what I say to Dr. Mann and others expressing deep concern over peer review: give up your data, methods and code freely and with a smile on your face. --Kevin Vranes, Science Policy, 18 February 2005

Mann's work doesn't meet that definition [of science], and those who use Mann's curve in their arguments are not making a scientific argument. One of Pournelle's Laws states "You can prove anything if you can make up your data." I will now add another Pournelle's Law: "You can prove anything if you can keep your algorithms secret." --Jerry Pournelle, 18 February 2005

The time has come to question the IPCC's status as the near-monopoly source of information and advice for its member governments. It is probably futile to propose reform of the present IPCC process. Like most bureaucracies, it has too much momentum and its institutional interests are too strong for anyone realistically to suppose that it can assimilate more diverse points of view, even if more scientists and economists were keen to join up. The rectitude and credibility of the IPCC could be best improved not through reform, but through competition. --Steven F. Hayward, The American Enterprise Institute, 15 February 2005   (

Jones, in turn, forwards the list of quotes to Mann Bradley and Hughes, authors of the study that McIntyre has critiqued.

From: Phil Jones <p.jones@uea.ac.uk>

To: mann@virginia.edu

Subject: Fwd: CCNet: PRESSURE GROWING ON CONTROVERSIAL RESEARCHER TO DISCLOSE SECRET DATA

Date: Mon Feb 21 16:28:32 2005

Cc: "raymond s. bradley" <rbradley@geo.umass.edu>, "Malcolm Hughes" <mhughes@ltrr.arizona.edu>

> *Mike, Ray and Malcolm,*
>
> *The skeptics seem to be building up a head of steam here ! Maybe we can use this to our advantage to get the series updated. The IPCC comes in for a lot of stick. Leave it to you to delete as appropriate !*
>
> *Cheers   Phil*
>
> *PS I'm getting hassled by a couple of people to release the CRU station temperature data.*
>
> *Don't any of you three tell anybody that the UK has a Freedom of Information Act !*

Jones apparently never considers the possibility that Wigley and Rind may be right when they point out the weakness of Mann's work. Instead, Jones sees this as an opportunity for funding to get the tree ring series updated. Later, Mann and others would argue that updating tree ring series is difficult work, a position McIntyre will prove wrong with his own tree ring field study. Finally, as Jones notes in his postscript, he is being bothered by a couple of people to release his CRU station temperature data, specifically Warwick Hughes and McIntyre. He is clearly worried that FOIA would be a problem for him. Inspired by Mann, who has made headlines by keeping his data secret, Jones finally answers Warwick Hughes' July 2004 request: on Feb 21 of 2005, the same day he reads the headlines generated by Mann's refusal to let others see hid data. With an example of stonewalling directly in front of him, Jones decides to follow Mann's example. As we saw before, he writes,

> *Subject: Re: WMO non respondo ... Even if WMO agrees, I will still not pass on the data. We have 25 or so years invested in the work. Why should I make the data available to you, when your aim is to try and find something wrong with it. ... Cheers Phil*

Jones, guided by Mann's example of animus toward McIntyre and others who question the work of climate scientists, and guided by Mann's relentless appeals to motive is transformed from a researcher who once promised to share data even in cases where it appeared there might be legal cause to withhold it, to a researcher who cares first about motive, and second about his own reputation and who in the end will use every legal and bureaucratic means to obstruct the release of temperature data and threaten its destruction.

CHAPTER FOUR: A PAPER IN PURGATORY

Cheat Sheet: What would normally be thought of as obscure academic infighting is actually becoming a tense struggle for control of the climate debate's agenda. The IPCC report deadline is fast approaching. If McIntyre's criticism of Michael Mann's work is the latest word in the scientific record, the IPCC will find it difficult to push its agenda forward. However, if a paper by Ammann and Wahl gets published before the deadline, then Keith Briffa, lead author of Chapter 6 of the IPCC report and member in good standing of The Team, will be able to dismiss McIntyre as having been refuted satisfactorily in the peer-reviewed literature.

Briffa's work on the First Order Draft (FOD) of Chapter 6 of the IPCC's AR4 continues under pressure from Overpeck. The paper by Ammann and Wahl, announced in a rare press release, continues to struggle through the publishing process and Overpeck is trying to manage the IPCC deadlines. The paper must be in press for Briffa to consider it, although Briffa already has a bootleg copy and Mann's interpretation of the unpublished paper.

Unfortunately for Overpeck, McIntyre has been appointed one of the expert reviewers for Chapter 6 and McIntyre is noting on his blog in real time the extraordinary efforts the Team is going through to get Ammann's paper into press. Eventually it does appear, but long after the IPCC deadline has passed. And in the end it doesn't show what it purports to show. But Chapter 6 has gone to press before anyone knows all the details.

In April 2005, McIntyre writes Jones, not using the FOIA as yet, and requests data. Not temperature data, but rather data about temperature proxies. Jones writes Mann and explains the state of his source code, perhaps the real reason why some scientists do not want to share code; simple embarrassment at the sorry state it is in:

> *From: Phil Jones <p.jones@uea.ac.uk>*
>
> *To: mann@virginia.edu*
>
> *Subject: Fwd: CCNet: DEBUNKING THE "DANGEROUS CLIMATE CHANGE" SCARE*
>
> *Date: Wed Apr 27 09:06:53 2005*
>
> Mike,
>
> *Presumably you've seen all this - the forwarded email from Tim. I got this email from McIntyre a few days ago. As far as I'm concerned he has the data - sent ages ago. I'll tell him this, but that's all - no code. If I can find it, it is likely to be hundreds of   lines of   uncommented fortran ! I recall the program did a lot more that just average the series.   I know why he can't replicate the results early on - it is because there was a variance   correction for fewer series.*

May 2005 starts with a rather unique event. In the run up to this, the mails reveal a team of scientists a bit perturbed by the fact that McIntyre has made press releases about his 2005 paper in GRL, which is definitely not how science has been traditionally communicated over the centuries. McIntyre is taking the fight to them on three fronts: in the journals, online, and in the press. The scientists take a page from his book and issue the following press release.

Boulder,CO May 10, 2005 Two new research papers submitted for review by Eugene R. Wahl (Alfred University) and Caspar M. Ammann (National Center for Atmospheric Research) reproduce the recently criticized Mann-Bradley-Hughes (MBH) climate field reconstruction, and invite researchers as well as the public to use the code for their own evaluation of the method.

Wahl and Ammann highlight the robustness of the MBH method against numerous modifications. Their detailed analyses (first presented publicly at the American Geophysical Union fall meeting, San Francisco, 2004 and at the American Association of Geographers annual meeting, Denver, 2005) reveal that the highly publicized criticisms of S. McIntyre and R. McKitrick against the "Hockey Stick" climate field reconstruction of MBH (Nature, 1998) are unfounded.

Conclusions of temperatures during the 15th century rivaling the late 20th century climate are found to be without statistical and climatological merit, as is the alleged identification of a fundamental flaw that would significantly bias the MBH climate reconstruction towards a hockey stick shape. Wahl and Ammann address each criticism raised by McIntyre and McKitrick (Energy and Environment 2003, 2005 and Geophysical Research Letters 2005) and find that: "High 15th century temperatures are only achieved with statistical models that don't pass validation tests, and thus lack both statistical and climatological meaning. Using mean 20th century climatology would provide a more successful reconstruction than the results put forth by McIntyre and McKitrick.

"If Principle Component analysis of North-American tree ring data is applied appropriately (i.e., with retention of all the relevant climate information contained in the tree ring data), climate reconstructions very similar to MBH are achieved, independent of the reference period applied. This is confirmed by an analysis leading to the same result if all individual tree ring series themselves are included. Code for Reconstruction based on Mann-Bradley-Hughes: http://www.cgd.ucar.edu/ccr/Ammannn/millennium/MBH_reevaluation.html "Real Climate" elaboration on various issues: http://www.realclimate.org Contact information: "Caspar Ammann, National Center for Atmospheric Research: 303-497-1705, Ammannn@ucar.edu ",Eugene Wahl, Alfred University: 607-871-2604 wahle@lafred.edu

The context surrounding the publication of these two papers is key. The writers of Chapter 6 of AR4 rely heavily on the work of Mann and his 98 paper which has been criticized by McIntyre in his publications, notably the 2003 paper and the 2005 paper in GRL. Since the findings of Chapter 6 are required to faithfully represent the scientific record, they must come to terms with McIntyre's criticism. The only tools they have are published papers. MBH 98 was published in peer reviewed literature and MM05 was now published in peer reviewed literature as well. MM05 cannot go unanswered otherwise the author of Chapter 6, Briffa, will be duty bound to take notice of it. Briffa, as other mails reveal, has some issues with Mann's absolute confidence that it was not warmer 1,000 years ago and so the existence of MM05 could provide him with the citations he would need to back up his less certain position.

Jones, who now views the struggle with McIntyre in epic (or at least Hollywood-ish) terms –the Empire Strikes Back—writes to Mann on the heels of the press release:

From: Phil Jones <p.jones@uea.ac.uk>

To: "Michael E. Mann" <mann@virginia.edu>

Subject: Empire Strikes Back - return of proper science !

Date: Fri May 20 13:45:26 2005

> *Mike,*
>
> *Just reviewed Caspar's paper with Wahl for Climatic Change. Looks pretty good. Almost reproduced your series and shows where MM have gone wrong. Should keep them quiet for a while. Also they release all the data and the R software. Presume you know all about this. Should make Keith's life in Ch 6 easy !*

The full history of the publication of these papers, and the final rejection of one of them is closely covered by Bishop Hill at his eponymous weblog, in a post called "Caspar and the Jesus Paper" a wry reference to the resurrection of the Ammann paper. (We should note that Bishop Hill has also published a book that includes his examination of Climategate, as well as the story of Caspar and the Jesus Paper.) Ammann and Wahl is an important paper primarily because it tried to protect an icon of climate science, Michael Mann's hockey stick, a graph which has taken on the status of a holy relic in climate science, almost like the Shroud of Turin, a comparison we make while noting that Turin's Shroud has been more or less thoroughly debunked. Overpeck has great hopes for Briffa's chapter and communicates that to Briffa, indicating that he hopes Briffa has something that will be used in the more visible sections such as the TS or Technical Summary:

From: Jonathan Overpeck <jto@u.arizona.edu>

To: Keith Briffa <k.briffa@uea.ac.uk>

Subject: IPCC - your section

Date: Mon, 23 May 2005 22:46:11 -0700

Hi Keith - thanks again for the help in Beijing. We hope you found a fabulous clay pot or at least some good views of China. We know it's going to be extra hard on you to get everything done on time, but we're hoping you can more-or-less stick to the schedule we just sent around. Your section is going to be the big one, and we need to make sure we have as much review and polishing as possible. If we don't we (especially you) will pay heavily at FOD [First Order Draft] review time. Lots of work now saves even more work later. Or so the real veterans tell us. Lastly, we wanted you to know that we can probably win another page or two (total, including figs and refs) if you end up needing it. Susan [Solomon] didn't promise this, but she gave us the feeling that we could get it if we ask - but probably only for your section, and maybe an extra page for general refs (although we're not going to mention this to the others, since we're not sure we can get it). Note that some of the methodological parts of your sections should go into supplemental material - this has to be written just as carefully, but it gives you another space buffer. All this means you can do a good job on figures, rather than the bare minimum. We're hoping you guys can generate something compelling enough for the TS and SPM - something that will replace the hockey-stick with something even more compelling. Anyhow, thanks in advance for what is most likely not going to be your number 1 summer to remember. That said, what we produce should provide real satisfaction. Best, Peck and Eystein

For people who ever questioned the importance of the hockey stick graphic and the pressure put on Briffa to quash his doubts about Mann's work, this mail should clarify the matter. It also clarifies the importance of "hiding the decline" or truncating the Briffa reconstruction of past temperatures. Hiding the decline does not mean hiding a decline in temperatures. It means concealing the fact tht the tree ring records show a decline in temperatures where the thermometers show an increase. What's at stake is the ability to trust the tree rings.

On June 28th[7th] Overpeck raises a concern with The Team. The question has been raised about the propriety of having authors, such as Briffa, review their own work. Can they remain objective? He advises them

> *Also, please note that in the US, the US Congress is questioning whether it is ethical for IPCC authors to be using the IPCC to champion their own work/opinions. Obviously, this is wrong and scary, but if our goal is to get policy makers (liberal and conservative alike) to take our chapter seriously, it will only hurt our effort if we cite too many of our own papers (perception is often reality). PLEASE do not cite anything that is not absolutely needed, and please do not cite your papers unless they are absolutely needed. Common sense, but it isn't happening. Please be more critical with your citations so we save needed space, and also so we don't get perceived as self serving or worse.*

In the end the team and Overpeck will do far more than this: they will violate the published deadline cutoffs for a wide assortment of papers, papers that come almost exclusively from a tight knit group of authors. They will discuss strategies to have editors removed from journals, interfere with the review process and keep critical papers out of publication. They will create a canon, a body of scientific literature that fits the story they want to tell to policy makers. Sadly, this body of science has not advanced our understanding of how and why climate has changed over the past 2,000 years. Rather, it has reduced trust not only in the climate record, but in the process science uses to investigate these issues.

The Race against the clock

The IPCC had determined a series of deadlines for material that the writing teams could cite or consider in writing their chapters. Those deadlines, as detailed on Climate Audit were as follows:

> **Deadlines for literature cited in the Working Group I Fourth Assessment Report**
>
> **The Working Group I Fourth Assessment Report (WG1-AR4) must be clearly based on the peer reviewed literature, noting that this should also include non-English language publications. Although there is provision in IPCC rules for citing unpublished or non peer-reviewed sources, the use of such material should be limited in WG I, as explained in more detail below.**

> The deadlines for publication of peer reviewed papers that can be cited in the WG I Fourth Assessment Report (AR4) are determined by the IPCC rules for the review process together with some practical considerations regarding the time to submit and review a journal paper. The constraints can be summarized as follows:
>
> When authors start to write the first draft for the AR4 they should have final or draft material for all work that is being cited in front of them. In practice this means that by May 2005, papers cited need to be either published or available to LAs [Lead Authors] in the form of a reasonably accurate draft of what is expected to be the final publication.
>
> When the first draft [FOD] of the AR4 is sent out for expert review, copies of any cited papers not already published must be made available to reviewers on request. This means that LAs need to ensure that drafts of any such papers are sent to the TSU before or at the same time as the chapter drafts, for which the absolute deadline is August 12. Such drafts will be made available to reviewers in confidence solely for the purposes of the review and are not to be copied, cited or re-distributed by them.
>
> When the second draft of the AR4 is written authors need to be sure that any cited paper that is not yet published will actually appear in the literature, is correctly referenced, and will not be subsequently modified (except perhaps for copy editing). In practice this means that by December 2005, papers cited need to be either published or "in press".

This is a critical deadline that Overpeck and his team are well aware of. In fact, the team will approach the editor of the journal Climatic Change, Professor Steven Schneider of Stanford University (who in any event has been included in the address list of many of the Climategate emails), and get him to invent a new category for papers: "provisionally accepted." They know about the deadlines. They know Briffa needs the paper in press in December so they work their friends to game the system.

So, by late February, Briffa must assure the TSU that the paper is in press and a preprint must be provided. The expert review teams which include McIntyre cannot effectively review the chapter unless they have access to the published literature that the writing team is citing.

> The above constraints are necessary because the IPCC assessment process is under intense scrutiny and we have an obligation to ensure that the literature is reported accurately and in a balanced way that is fair to the science community, the review process, and our final policymaker audience.

And so in its own guidelines the IPCC holds that abiding by these constraints is necessary to fundamental fairness. Overpeck, aware of the deadlines and requirements for the first order draft sends a mail to Wahl, co-author with Ammann of the paper that is supposed to handle the MM05 issues, inquiring about the status of the paper:

From: Jonathan Overpeck <jto@u.arizona.edu>

To: Caspar Ammannn <Ammannn@ucar.edu>

Subject: Re: What's up with your paper with Eugene?

Date: Fri, 1 Jul 2005 12:46:59 -0600

Hi Caspar and Gene - Thanks. I look forward to hearing how things go - if the paper is in press by the first week of August, we'll cite it in the Chapter 6 of the FOD, but otherwise I guess it'll have to wait - that's ok too. But.keep us posted (and send revised preprint when possible). Thanks! Peck

Overpeck, of course, has a copy of the published deadlines. For literature to be considered it must be published by certain dates or be in the hands of reviewers in a form that will be published. The exact deadlines that Peck and Briffa are working toward is captured in this IPCC document

IPCC Working Group I

> **Schedule for Fourth Assessment Report 2005 Meeting of the TS/SPM writing team December 16, Christchurch, New Zealand** *Note. Literature to be cited will need to be published or in press by this time.* **2006 Jan Preparation of the second draft Feb Preparation of the second draft Annotated responses to all comments on the first draft need to be completed and sent to the TSU by CLAs by** *February 3* **Material from second draft Chapters to be used in the Technical Summary to be sent to the TSU by** *February 10* **March: Second order draft of chapters to be submitted by CLAs to TSU by** *March 3* **Copies of literature not available through normal library sources should be sent to the TSU so they can be made available to reviewers if requested. to TSU by March 3.**

Jones sends an e mail to John Christy, a well known, and frequently published, skeptical scientist, and outlines his view of things, in particular an odd view on climate change, arguing in effect that he hopes that nothing will be done so that the science and Phil Jones can be proven correct. Jones does not want to release his data because he is afraid of being shown wrong and here he contemplates hoping the planet gets warmer to prove that he is correct: Also, important to note is that Jones' attitude toward the deadlines of the IPCC, which were instituted to insure fairness. In Jones's mind they are silly. In Jones's mind he wants to decide what is important.

> *What will be interesting is to see how IPCC pans out, as we've been told we can't use any article that hasn't been submitted by May 31. This date isn't binding, but Aug 12 is a little more as this is when we must submit our next draft - the one everybody will be able to get access to and comment upon. The science isn't going to stop from now until AR4 comes out in early 2007, so we are going to have to add in relevant new and important papers. I hope it is up to us to decide what is important and new. So, unless you get something to me soon, it won't be in this version. It shouldn't matter though, as it will be ridiculous to keep later drafts without it. We will be open to criticism though with what we do add in subsequent drafts. Someone is going to check the final version and the Aug 12 draft. This is partly why I've sent you the rest of this email. IPCC, me and whoever will get accused of being political, whatever we do. As you know, I'm not political. If anything, I would like to see the climate change happen, so the science could be proved right, regardless of the consequences. This isn't being political, it is being selfish. Cheers*

Mann's concern about skeptics is that they are directed by corporate interests, the profit motive or by political motives. Jones argues that he is not political, but rather selfish. He wants 'his' science, his gut feeling, proved right.

The importance of the Ammann Wahl paper is clear to Jones. He explains to Simon Tett of MET on July 29 what the IPCC will find. Of course this is prior to the Ammann paper actually being through peer review and changes:

> *Simon,*
>
> *If you go to this web page [1]http://www.ucar.edu/news/releases/2005/Ammannn.shtml You can click on a re-evaluation of MBH, which leads to a paper submitted to Climatic Change. This shows that MBH can be reproduced. The R-code to do this can be accessed and eventually the data - once the paper has been accepted. IPCC will likely conclude that all MM arguments are wrong and have been answered in papers that have either come out or will soon.*

The Cat Bird Seat

On Sept 20th, McIntyre, acting as an outside reviewer of IPCC AR4 Chapter 6, notes that the draft cites two papers, one by Wilson and the other by Briffa, which are not available to him. As a reviewer of Chapter 6, McIntyre sits in the cat bird seat. He, like all the parties working on the document, is well aware of the deadlines and procedures. Overpeck writes to the Team about McIntyre's request for the papers, which should not be at all controversial, given his role.

> *From: Jonathan Overpeck <jto@u.arizona.edu>*
>
> *To: ÿyvind Paasche <oyvind.paasche@bjerknes.uib.no>*
>
> *Subject: Re: Fwd: Re: [Fwd: Re: Chapter 6 - Submitted Papers]*
>
> *Date: Sat, 24 Sep 2005 22:10:05 -0600*
>
> *Hi all - let's see what Keith/Tim say about both papers. Eystein - can you call them on Monday if we haven't heard from them. If they don't have one or both of the papers, then we should ask Martin to delete from the chapter - Eystein, feel free to do this as soon as you get feedback from Keith/Tim. Mysterious... Thanks, Peck*

Papers must either be published or must be available to reviewers in a form that will eventually be published. The chain of evidence must be clear. As of November 30th the Wahl and Ammann paper which will be used to handle MM05 is still in review and has not been accepted by the journal Climate Change. Behind the scenes they work with Climate Change editor Stephen H. Schneider to "invent" a new category for papers: "provisionally accepted."

2005 comes to a close. The IPCC December 16 deadline passes and the Ammann Wahl paper has yet to be accepted, much less published: Overpeck:

From: Jonathan Overpeck <jto@u.arizona.edu>

To: Eystein Jansen <Eystein.Jansen@geo.uib.no>

Subject: Re: Fwd: RE: Wahl-Ammannn paper on MBH-MM issues

Date: Mon, 23 Jan 2006 13:47:30 -0700

> Hi all - I'm betting that "provisional acceptance" is not good enough for inclusion in the Second Order draft, but based on what Gene has said, he should have formal acceptance soon - we really need that. Can you give us a read on when you'll have it Gene? Best make this a top priority, or we'll have to leave your important work out of the chapter. Many thanks!! Peck

The games that The Team are playing with deadlines is not lost on McIntyre, attentive as always to procedure. Eystein, who is reading Climate Audit and fears being caught violating the regulations explains, referring to the "new" category of provisional acceptance:

> Hi Peck, I assume a provisional acceptance is OK by IPCC rules? The timing of these matters are being followed closely by McIntyre (see: http://www.climateaudit.org/?p=503) and we cannot afford to being caught doing anything that is not within the regulations. Thus need to consult with martin and Susan on this (see also last mail from Melinda). Cheers, Eystein

McIntyre writes in Climate Audit on Jan 17th

> **Ammann Chronology**
>
> **I've just noticed at the UCAR website that Ammann and Wahl now say that their CC [Climate Change] re-submission was "provisionally accepted" on Dec 12. I have no information on what a "provisional acceptance" means, but it's certainly a coincidence that the "provisional acceptance" occurred only 3 days after GRL agreed to send their previously rejected GRL comment out for review, together with an expected reply from us. This is a second coincidence: they re-submitted to CC on Sept 27, a few days after they were allowed to re-submit to GRL on Sept 25 after getting their editor changed at GRL. Maybe it's just a coincidence; but perhaps CC acceptance is contingent on their GRL submission not being rejected another time.**

As noted earlier, Mann and others had issues with the editor, Saiers, at GRL because he had accepted a paper from McIntyre. By this time in 2005 that editor had been replaced with one more friendly to their point of view, James Famigliette.

The editor at Climatic Change, the prestigious Stanford climate scientist Stephen H. Schneider, was working with Ammann and Wahl to try to get the paper into a form that can be released to reviewers, which means that it must be in the same form as final publication. At issue is the presentation of a key statistic, r^2, which will tell readers if the result has any significance or not. Over the course of the history of this paper McIntyre will push to have this statistic revealed and he will be fought at every step by the authors. In Mann's original 1998 paper Mann had not reported the r^2 statistic. McIntyre had replicated Mann's work and showed that r^2 was equal to zero, that is Mann's reconstruction had no skill (a way of saying it wasn't useful). As Ammann's work aimed at answering MM05, reporting r^2 was key to blunting McIntyre's attack on MBH98.

The paper is being delayed while Wahl and Ammann deal with this issue. Wahl and Susan Solomon are both aware of the importance of this issue as Wahl writes:

> *From: "Wahl, Eugene R" <wahle@alfred.edu>*
>
> *To: "Jonathan Overpeck" <jto@u.arizona.edu>*
>
> *Hello Jonathan and Keith:*
>
> *I'm not sure that I ever sent you the updated Wahl-Ammannn paper that was the basis for Steve's [Schneider] provisional acceptance. Here it is. As is, it contains a long appendix (# 1) on issues with interannual statistics of merit for validation, which was not in the version I had sent you earlier in the year. All the main results and conclusions are the same. Caspar and I are also now responding to Steve's final requests, based on independent re-review. This is primarily to address publishing Pearson's r^2 and CE calculations for verification, which Steve and the reviewer reason should be done to get the conversation off the topic of us choosing not to report these measures, and onto the science itself. We explain thoroughly in the appendix I mention above why we feel these (and other interannual-only) measures of merit are not of much use for verification in the MBH context, so that the fact we are reporting them is contextualized appropriately. IN FACT, we will be going farther than that and will be bringing this material currently in an appendix into the main text, based on the reasoning below(quoted from another message) Caspar mentioned yesterday that he talked with Susan Solomon about this paper, and she did not see the appendix we had added concerning the issues about Pearson's r^2 etc. Based on this she therefore thought our text was weak in this area in relation to McIntyre's criticisms. Caspar thought, and I agree, that we need to bring this stuff OUT of the appendix and get it INTO the methods section, so that it won't be so easily missed!! We are working on this--which will include other material as well in the text proper. Also, we are going ahead with an even further-expanded discussion on the issues with r^2, which itself will probably become an appendix in the final text (it had been slated for publication as supplemental web-site material). This expanded discussion will go into additional reasoning (with graphics) concerning the basis for r^2 not being useful in this context.*

Briffa, working on the draft for the chapter continues to express concerns about overstating the certainty of the reconstructions or the skill the reconstructions have in accurately reproducing past temperatures. He writes:

> *From: Keith Briffa <k.briffa@uea.ac.uk>*
>
> *To: Jonathan Overpeck <jto@u.arizona.edu>,Eystein.Jansen@geo.uib.no*
>
> *Date: Fri Feb 3 14:31:09 2006*
>
> *Peck and Eystein we are having trouble to express the real message of the reconstructions - being scientifically sound in representing uncertainty, while still*

getting the crux of the information across clearly. It is not right to ignore uncertainty, but expressing this merely in an arbitrary way (and as a total range as before) allows the uncertainty to swamp the magnitude of the changes through time .

Overpeck and Wahl continue to discuss the status of the paper, especially the r^2 statistic. In order to meet the IPCC deadline of Dec 16th the editor at Climate Change tried to invent a new category "provisionally accepted" and now Wahl and Overpeck discuss how they can get the paper "in press" by the end of February, but the work is not yet done. And if it's not yet done then no reviewer can review it. But Schneider, friend and ally at the journal, is in play to help them meet the deadline.

From: Jonathan Overpeck <jto@u.arizona.edu>

To: "Wahl, Eugene R" <wahle@alfred.edu>

Subject: RE: Wahl-Ammannn paper and UAZ position

Date: Fri, 10 Feb 2006 12:05:44 -0700

Hi Eugene - this is good news... I hope. Please contact Steve[Schneider] and see if we will have "in press" status before the end of the month. He knows the drill, but also the downside of not being precise. Let me, Eystein and Keith know as soon as you know. Bit nuts right now, really appreciate your help.

thanks, peck

Hi Peck: Well, as I have understood it in our communications with Steve, final acceptance is equivalent to being in press for Climatic Change because it is a "journal of record". However, this would need to be confirmed to be quite sure. If that is the case, then in press is still possible by the end of the month. I think. Which would be best at this point, for me to write and ask Steve this, or would it be better for you to ask? I'm happy to do so, I just want to act in the most time-effective and appropriate way. I apologize for the fact that it is coming right down to the wire. The status right now is that I am waiting for final analytical results from Caspar re: Pearson's r and CE results on all the scenarios we have done. These results will go in an appendix table and I have to write a brief text to go with them for contextualization purposes--I already have in mind what I want to say. The entire rest of the document is essentially done. Steve turned around the change from "in review" to "provisionally accepted" within days last December after receiving back the final independent re-review (it had been due a month earlier), so I can imagine that he could potentially turn around the change from "provisional acceptance" to "full acceptance" similarly quickly. Please advise about who is best to contact Steve--and if me I will get on it today. Peace, Gene

And Overpeck knows full well the importance of this.

From: Jonathan Overpeck [mailto:jto@u.arizona.edu]

>Sent: Fri 2/10/2006 12:39 PM

>To: Wahl, Eugene R

>Cc: Eystein Jansen; Keith Briffa; t.osborn@uea.ac.uk

>Subject: RE: Wahl-Ammannn paper and UAZ position

Hi Gene - First the IPCC, then I'll send another email wrt UA Geography Based on your update (which is much appreciated), I'm not sure we'll be able to cite either in the SOD due at the end of this month (sections will have to be done this week, or earliest next week to meet this deadline). The rule is that we can't cite any papers not in press by end of Feb. From what you are saying, there isn't much chance for in press by the end of the month? If this is not true, please let me, Keith, Tim and Eystein know, and make sure you send the in press doc as soon as it is officially in press (as in you have written confirmation). We have to be careful on these issues. Thanks again, Peck

Wahl and Ammann press the editor at CC to get the paper in press, and Overpeck presses the publishing deadlines, Briffa feels the pressure from Mann and others to claim more certainty than he thinks the publishing record merits. Briffa makes points that McIntyre has made from the outset. First, that there is no real independence between the various studies that support the "hockey stick." Next that the various statistical methods used on the data are far from established and that not much has changed between the 3rd assessment and the 4th. Briffa is being pushed to sell more certainty than he believes is warranted.

Peck, you have to consider that since the TAR , there has been a lot of argument re "hockey stick" and the real independence of the inputs to most subsequent analyses is minimal. True, there have been many different techniques used to aggregate and scale data - but the efficacy of these is still far from established. We should be careful not to push the conclusions beyond what we can securely justify - and this is not much other than a confirmation of the general conclusions of the TAR . We must resist being pushed to present the results such that we will be accused of bias - hence no need to attack Moberg . Just need to show the "most likely" course of temperatures over the last 1300 years - which we do well I think. Strong confirmation of TAR is a good result, given that we discuss uncertainty and base it on more data. Let us not try to over egg the pudding. For what it worth , the above comments are my (honestly long considered) views - and I would not be happy to go further . Of course this discussion now needs to go to the wider Chapter authorship, but do not let Susan (or Mike) push you (us) beyond where we know is right.

On Feb 21, Wahl delivers a manuscript that he says shouldn't be passed around until it is in press:

From: "Wahl, Eugene R" <wahle@alfred.edu>

To: "Jonathan Overpeck" <jto@u.arizona.edu>

Subject: RE: Wahl and Ammannn Climatic Change article on MBH

Date: Tue, 21 Feb 2006 19:26:44 -0500

OK:

Here is the mss.[manuscript] Yes, fingers crossed. Note, this is not for general dissemination until actually "in press". The article is quite long, due to all the MM issues we address and the extensive discussions concerning use of validation measures we get into. As a first pass, the Abstract, Discussion, and Summary would be good places to start. Peace, Gene

On February 24th Wahl provides an update but notes that Schneider is still reviewing the paper. On the last day of the month as promised the editor delivers and calls the paper "in press."

From: "Wahl, Eugene R" <wahle@alfred.edu>

To: "Jonathan Overpeck" <jto@u.arizona.edu>

Subject: RE: Wahl Ammannn Climatic Change article on MBH/MM

Date: Tue, 28 Feb 2006 21:42:42 -0500

Hello all:

Good news this day. The Wahl-Ammannn paper also has been given fully accepted status today by Stephen Schneider. I copy his affirmation of this below, and after that his remark from earlier this month regarding this status being equivalent to "in press". I hope this meets the deadline of before March 1 for citation.

Peace, Gene

And in May the issue is still confused as the copy of the paper that is available to reviewers differs from the copy on the web. In short, Lead Author (LA) Briffa has been struggling against those who want more certainty in his chapter, principally Mann. And to bolster Mann's case Wahl, Ammann and Overpeck have been pressuring journals to accept a paper to meet IPCC deadlines.

In 2008, prior to the release of the Climategate files, McIntyre writes at Climate Audit about the strange sequence of events, and his suspicions are confirmed by the mails:

> **The first deadline cited above was May 10-13, 2005. On May 10, 2005, Ammann and Wahl submitted Wahl and Ammann 200x to Climate Change and Ammann and Wahl 200x to GRL -see** here **– and issued a press release on May 11, 2005** here **. The submission date on the** published version **of Wahl and Ammann 2007 is May 11, 2005.**
>
> **The second deadline was December 13-16, 2005 and "literature to be cited will need to be published or in press by this time". The UCAR website states that Wahl and**

Ammannn had been "Provisionally Accepted" on Dec. 12, 2005 – on the eve of the expiry of the second IPCC deadline. Has anyone ever seen a paper report the date of its "Provisional Acceptance"? Clearly "provisional acceptance", whatever it is, is not the same thing as being "published or in print". Wahl and Ammann had clearly missed an important deadline – a deadline that Judith Curry and others were adhering to.

The end of February 2006 was a drop-dead date by which "the TSU must hold final preprint copies of any unpublished papers that are cited in order that these can be made available to reviewers. This means that by late-February 2006 if LAs cannot assure us that a paper is in press and provide a preprint we will ask them to remove any reference to it." The UCAR website states that Wahl and Ammann was supposedly "accepted for publication" on Feb 28, 2006, this third date, like the other two, being on the eve of an IPCC deadline. So there is an incontrovertible pattern.

The timing of the Climatic Change decision was very curious, to say the least. GRL turn times are very short. The GRL decision on the contingent GRL submission would be available in 2 weeks; within the four corners of journal administration, there was no reason for Climatic Change to make a hurried decision ahead of GRL – especially given that Schneider had already been burnt once by Ammann on his GRL submission. Why wouldn't Schneider just wait a couple of weeks? Or is it possible that UCAR/Ammannn, with an eye on IPCC deadlines, put pressure on Schneider to "accept" the article before the end of February? Sure seems likely.

What McIntyre suspected before seeing the Climategate mails was in fact true. Schneider "knew the drill." He was called on to invent a new category "provisional acceptance" to meet a mid December deadline and then pressed again to get the paper "in press" by the end of February. The Ammann and Wahl paper was actually not published until 2007. And when it was finally published it contains references to papers written in 2006, papers published after the Feb 28th date.

Jones, who knew of the deception, suggested another subterfuge:

From: Phil Jones [mailto:p.jones@uea.ac.uk]

Sent: Wednesday, September 12, 2007 11:30 AM

To: Wahl, Eugene R; Caspar Ammannn

Subject: Wahl/Ammannn

Gene/Caspar,

Good to see these two out. Wahl/Ammannn doesn't appear to be in CC's [Climate Change] online first, but comes up if you search. You likely know that McIntyre will check this one to make sure it hasn't changed since the IPCC close-off date July 2006!

Hard copies of the WG1 report from CUP have arrived here today. Ammannn/Wahl - try and change the Received date! Don't give those skeptics something to amuse themselves with. Cheers, Phil

And Wahl responds:

> *From: Wahl, Eugene R*
>
> *Sent: Wednesday, September 12, 2007 6:44 PM*
>
> *To: 'Phil Jones'; Caspar Ammannn*
>
> *Subject: RE: Wahl/Ammannn*
>
> *Hi Phil:*
>
> *There were inevitably a few things that needed to be changed in the final version of the WA [Wahl & Ammann used in chapter 6] paper, such as the reference to the GRL paper that was not published (replaced by the AW paper here), two or three additional pointers to the AW paper, changed references of a Mann/Rutherford/Wahl/Ammannn paper from 2005 to 2007, and a some other very minor grammatical/structural things. I tried to keep all of this to the barest minimum possible, while still providing a good reference structure. I imagine that MM will make the biggest issue about the very existence of the AW paper, and then the referencing of it in WA; but that was simply something we could not do without, and indeed AW does a good job of contextualizing the whole matter.*
>
> *Steve Schneider seemed well satisfied with the entire matter, including its intellectual defensibility (sp?) and I think his confidence is warranted. That said, any other thoughts/musings you have are quite welcome.*
>
> *Peace, Gene*

The Team had what it needed—a peer-reviewed paper that tried to answer McIntyre's criticism of Mann that they could cite in the IPCC report. They had won this battle, but at quite a cost. Their tactics had been reported, and word was getting out about how they did business. And McIntyre, in using the Freedom of Information Act to find out who said what to who, would find out a lot more.

Perception is often reality

In July of 2005, 2 months after the Ammann and Wahl paper was announced, long before the unanticipated publication delays, Overpeck wrote:

but if our goal is to get policy makers (liberal and conservative alike) to take our chapter seriously, it will only hurt our effort if we cite too many of our own papers (perception is often reality). PLEASE do not cite anything that is not absolutely needed, and please do not cite your papers unless they are absolutely needed

This set of instructions seems to have been forgotten. Not only did the team cite themselves extensively, they bent the rules for their own papers. In 2008 McIntyre details the actions taken by the Chapter 6 team with regards to accepting papers past the deadlines. It's a damning indictment of the objectivity of the people involved. McIntyre posts the following and details the special treatment this paper received

> AR 4 Chapter 6 – "In Press" and "Accepted" Articles
>
> I examined the "In Press" and "Accepted" citations in IPCC AR4 Second Draft Chapter 6 to verify whether Wahl and Ammannn 200x had received unusual and special treatment. It definitely did; it's surprising how much so. There was also a very interesting tendency for IPCC Authors to bend the rules in their own favor.
>
> IPCC Chapter 6 contained 21 bibliographic citations to "In Press" or "Accepted" articles, of which 20 were to journal articles and one (Kaspar and Cubasch was in a book collection). In the table shown below, I've collated the Second Draft citations to these 20 journal articles, plus the citation to Osborn and Briffa 2006, which also missed the December 2005 acceptance deadline.
>
> Of these 21 citations in the Second Draft to "In Press" or "Accepted", all but 2 are to articles in which an IPCC Contributing Author is a coauthor. The only two such articles which were by third parties are Royer (2006) and Shin et al (2006).
>
> Of these 21 articles, only 5 had been accepted by the December 2005 deadline, two of which were the two third party articles. Thus 16 articles – all by IPCC Contributing Authors – failed to meet an important IPCC Publication Deadline and were accepted after the December 2005 final date, including, of course, Wahl and Ammannn 2007. There were zero citations of third party articles that failed to meet the IPCC deadline. Obviously these rules were preferentially bent in favor of IPCC Contributing Authors.
>
> Most of these 21 "In Press" and "Accepted" articles were published during the next few months. By early August, when the Chapter Author Replies were due, only three articles had not actually been published – de Vernal et al 2006, Tett et al 2006 and, needless to say, Wahl and Ammannn 2007. De Vernal et al 2006 was published on Aug 21, 2006 and Tett et al 2006 was published on 22 September 2006. In order to comply with the July 2006 exemption for unpublished articles, iron-clad certificates were supposed to be delivered to TSU. Both De Vernal et al 2006 and Tett et al 2006 would have been able to deliver such certificates. Only one journal citation in the entire Chapter 6 corpus remained unpublished 6 weeks after the Chapter Author Reply date – Wahl and Ammannn 2007. Did they deliver the required certificate to IPCC TSU? Inquiring minds want to know.
>
> Wahl and Ammannn 200x wasn't even published in 2006. In fact, it wasn't even published in the first half of 2007. It was published on August 31, 2007 , over a year later.
>
> It's not as though IPCC wasn't on explicit notice of the problem with Wahl and Ammannn 200x. Two alert reviewers had identified its compliance problems and Chapter 6 Authors, and presumably the TSU, had both the obligation and opportunity to deal with non-compliance by Wahl and Ammannn.

> **Instead of dealing with Ammannn's non-compliance, IPCC Chapter Authors, one of whom (Otto-Bliesner) had been Ammannn's direct supervisor, falsely stated that Wahl and Ammannn 200x complied with the "new/revised/current guidelines". Review Editor Mitchell was supposed to check the Replies. What steps did Mitchell take to verify that these Replies were true? Doesn't look like he did anything.**

This episode in 2006 will prove critical in the years to come as one of the targets of McIntyre's FOIA requests will be the reviewer's comments. By issuing an FOIA for his own comments on Ar4 he will show in effect is that Ammann was providing information to Briffa outside of the IPCC review process. McIntyre will be aided in this effort by Holland who will request the correspondence of Briffa, Osborne and Ammann.

Briffa was receiving communication outside the IPCC process. This fact does not, of course, make the "science" wrong. What it illustrates is that the scientists were driven to subvert the process to deliver a particular message. It invites objective readers to put the results of Chapter 6 into question.

The seriousness of subverting the IPCC process is captured best by the following series of mails. In May of 2006 Neil Roberts, sends Overpeck the following mail:

> *Dear Jonathan   Please excuse me for writing direct, but Keith Briffa suggested it would be simplest. I have looked through the draft chapter 6 and find it an impressive document. However, bullet 4 on page 6.2, starting "global mean cooling and warming....." strikes me as   incorrect and misleading.   …   In short, this particular bullet seems in need of critical   reassessment before the definitive version of the next IPCC reprot   emerges.   Thanks in anticipation and best regards   Neil*

Briffa has suggested that Roberts write directly. Overpeck, who as we have seen has been pushing Ammann07 to hit a deadline, writes back. The IPCC has strict rules. All comments must come through the review process:

> *From: Jonathan Overpeck <jto@u.arizona.edu> To: "Neil Roberts" <C.N.Roberts@xxxxxxxx.xxx>   Subject: Re: ipcc chapter 6 draft Date: Thu, 18 May 2006 15:58:25 -0600   Cc: Keith Briffa <k.briffa@xxxxxxxx.xxx>, Eystein Jansen*
>
> *Hi Neil - Thanks for your interest in providing feedback on the draft   chap 6 Second Order Draft. Since the IPCC has very strict rules about   all this, I'm going to ask them (the IPCC) to send you an official   invitation to review, along with the process   formal, but highly   efficient - to follow. If you could send your comments in that way it   would be a great help. We've been asked to keep everything squeaky   clean, and not to get comments informally.   Thanks! Peck*

Later Briffa, in an apparent violation of IPCC procedure, sends copies of the comments to Wahl and asks for comments back. Comments that would be outside the process:

From: Keith Briffa [mailto:k.briffa@uea.ac.uk]

Sent: Tue 7/18/2006 10:20 AM

To: Wahl, Eugene R

Subject: confidential

Gene

I am taking the liberty (confidentially) to send you a copy of the reviewers comments (please keep these to yourself) of the last IPCC draft chapter. I am concerned that I am not as objective as perhaps I should be and would appreciate your take on the comments from number 6-737 onwards , that relate to your reassessment of the Mann et al work. I have to consider whether the current text is fair or whether I should change things in the light of the sceptic [MM05] comments. In practise this brief version has evolved and there is little scope for additional text , but I must put on record responses to these comments - any confidential help , opinions are appreciated . I have only days now to complete this revision and response. note that the sub heading 6.6 the last 2000 years is page 27 line35 on the original (commented) draft.
Cheers

Keith

And Wahl responds. It should be noted that his guess at when Ammann07 will finally be in press (2/28/06) will be off by more than a year.

From: "Wahl, Eugene R" <wahle@alfred.edu>

To: "Keith Briffa" <k.briffa@uea.ac.uk>

Subject: RE: confidential

Date: Sat, 12 Aug 2006 13:02:44 -0400

Hi Keith:

Thanks so much for the chance to look over this section. I think the long section you added on pp 6-5 and 6-6 reads well, and makes good sense according to what I know. Indeed, reading the whole section is a good review for me!

I suggested addition of a phrase in lines 32-33 on page 6-3 regarding MM 2003 and analysis of it by Wahl-Ammannn 2006. I also suggest a (logically useful) change from singular to plural in line 42 of that page. The changes are in RED/BOLD font.

[I should note that AW 2006 is still in "in press" status, and its exact publication date will be affected by publication of an editorial designed to go with it that Caspar and I are submitting this weekend. Thus I cannot say it is certain this article will come out in 2006, but its final acceptance for publication as of 2/28/06 remains completely solid.]

Later, as we will see, Holland requests under FOIA, that communication between Wahl and Briffa be released to him. Jones will direct Briffa to say he only received communication under the IPCC process. Then Jones will direct Mann to have Wahl delete his emails. Incidents like this crystallize the concerns raised by Climategate. It's clear from the record that Ammann and Wahl 2007 was given special treatment. That fact does not go to the basic science involved; that is, merely because Ammann and Wahl received special treatment does not invalidate their findings. What it does go to is the trust people can put in the IPCC process and the people involved. As Overpeck himself noted, perception is often reality.

CHAPTER FIVE. FREE THE DATA; FREE THE CODE

Cheat Sheet: This is the period of time where a Congressional committee asks the National Academy of Sciences and noted statistician Dr. Edward Wegman to investigate Michael Mann's Hockey Stick calculations and Steve McIntyre's criticism. The NAS and Wegman would support McIntyre on every point he made, and Wegman in particular was trenchant about his comments regarding Mann and his team, going so far as to almost predict the events of Climategate, due to the close-knit, almost incestuous inter-relationships between members of The Team, who served as co-authors, reviewers, editors and promoters of each others' work.

This chapter should serve to put one of the accusations in our preface into context—when we speak of frustrating the intent and proper practice of the UK's Freedom of Information Act, we are not speaking of an isolated case, or one instance of an intemperate outburst by Phil Jones. It was a coordinated strategy, used consistently, involving several members of The Team and apparently a cooperative or co-opted Freedom of Information officer.

Phil Jones' fear that McIntyre will discover FOIA regulations comes true. The discussion on UHI at Climate Audit leads naturally to a discussion of Jones' seminal 1990 paper on urban heat islands. Jones' 2005 refusal to provide data to Warwick Hughes is discussed and Climate Audit regular Willis Eschenbach, who is eager to check Jones' work, submits the first known related FOIA request to Jones. McIntyre and others follow suit and they try to get a straight answer from CRU, who seem intent upon misunderstanding simple requests. At first they seem to be caught by surprise and reply with several self-contradictory reasons to refuse to surrender any data. But with a lot of emailed communications between members of The Team, they get their story straight. In the end CRU agrees to release some of the data requested for a paper that was published in 1990.

During the summer of 2006 McIntyre was spending his time writing posts on various statistical matters related to the reconstruction of past climates, as well as some posts on the National Academy of Science (NAS) panel which had been investigating Mann's work and a congressional committee hearing.

In an op-ed piece on the NAS panel McIntyre wrote,

> "In February, 2006, the NAS appointed a panel of 12 eminent academics involved in climate science but not directly involved in the temperature reconstructions of the past 1,000 years. They were not an entirely "independent" panel, as some were

occasional co-authors with the Hockey Stick authors. But even this limited independence was a major departure from procedures of the IPCC, which permits authors actively involved in scientific controversy to summarize the research -- even if they end up acting as reviewers of their own work! In March, 2006, the NAS panel held meetings in Washington at which we made a presentation (along with Mann and seven other scientists in the field).On July 6, the panel issued a 155-page report, which managed the delicate feat of accepting virtually all the criticisms of the Hockey Stick while still saying polite things about it. A European climate scientist, who understood the balancing act, wrote us afterwards to point out it was the most severe criticism of the Hockey Stick nowadays possible At the NAS panel, we said that Mann's principal components were biased toward producing hockey stick-shaped series; the NAS agreed. We said that bristlecones were not a reliable temperature proxy; the NAS agreed and said they should be "avoided." We said that Mann's reconstruction failed important verification tests; the NAS agreed. We said that more than one test statistic should be reported when assessing statistical validity; the NAS agreed. We said that current methods underestimated the inherent uncertainty; the NAS agreed. On and on. On no occasion was any claim of ours refuted."

As McIntyre continued to post articles on data access and its importance in the role of science, he cited one of the darlings of climate science, Naomi Oreskes, a professor of History and Science Studies at UC San Diego. In 2004, Oreskes had published an essay titled, "Beyond the Ivory Tower: The Scientific Consensus on Climate Change", which examined the abstracts of 928 scientific papers on climate change and came up with the conclusion that 75% supported the consensus view of the issue, while none directly disputed it. (Later research revealed that there were over 12,000 papers she could have included in the study, including many that directly disputed elements of global warming theory. After another scientist published a paper that found some errors in her essay, Oreskes issued a partial retraction of the numbers she had used, while still maintaining, probably correctly, that the vast majority of climate scientists concur with the proposition that anthropogenic emissions of greenhouse gases are a significant contributor to global warming.) Oreskes wrote:

Jul 6, 2006

Trust, but verify! This is what editors ask for, and what readers expect, from reviewers of technical articles. As a reviewer, I am growing concerned with the level of trust requested by authors of submitted manuscripts, and the frequent lack of verifiable data and methods. Negative reports in the press [e.g., New York Times, 2005] attest to the worst-case outcomes of such shortcomings ....

Where scientific findings are based on computational analyses, documentation of computer model methods and analyses ought to be a required element of publication. The trust of the public in scientists and our methods depends upon this.

It occurred to McIntyre that perhaps he could get the data he wanted from the NAS itself which had undertaken the review of the climate reconstruction disputes. McIntyre had requested data from Jones. Jones had refused. Now the NAS panel had reviewed some of the science. Did they take the utterly common first step in any review? Did they request the underlying data and methods, or did they just read papers?

> **McIntyre wrote:**
>
> **The NAS panel relied heavily on articles using unarchived data and I plan to send a letter to Ralph Cicerone, President of NAS, asking him to directly request unarchived data from the various authors. Here is a draft. ……..**
>
> Dear Dr Cicerone, I am currently in the process of analyzing the interesting report of the National Academies of Science panel on Surface Temperature Reconstructions. The panel has, to a considerable extent, relied on studies for which data is unarchived and/or methods are insufficiently described to enable replication. In most cases, I have tried unsuccessfully to obtain this information. Now that the NAS has relied on this information, I request that you request that the information listed below be archived or, failing that, provided to you so that you can forward the information to me.
>
> I apologize for involving you in this process. However, I have made diligent and unsuccessful efforts to obtain this information. I believe that a request from you would be appropriate given the recommendations of the NAS panel and might well be effective.
>
> Yours truly, SM
>
> (later…)
>
> APPENDIX…..
>
> 7. Phil Jones   The NAS panel used CRU temperature data as a reference point. Both von Storch and I quoted Phil Jones' notorious refusal to archive   a) supporting data, including station data;   b) detailed methodological information and/or source codes.

Sadly for McIntyre, Dr. Cicerone would not ask the scientists for their data. The NAS panel was set up to answer various questions posed by Congress. One of the questions it was supposed to address was the issue of replication. Could people replicate the work the scientists had claimed to perform in their papers? NAS refused to address this question in a forthright manner.

Meanwhile, for scientists working on Chapter 6 of AR4, the daily struggle with the document continued. Although it was being treated as a finished, peer-reviewed paper for the purposes of the IPCC AR4 report, they were still sort of working on it:

> *From: Tim Osborn <t.osborn@uea.ac.uk>*
>
> *To: Jonathan Overpeck <jto@u.arizona.edu>, Eystein Jansen <eystein.jansen@geo.uib.no>, Keith Briffa <k.briffa@uea.ac.uk>, ÿyvind Paasche <oyvind.paasche@bjerknes.uib.no>*

Subject: latest figures, captions and tables from Keith/Tim

Date: Mon Jul 31 14:31:24 2006

Dear all,

we have now updated the figure captions for our section and these are attached as a PDF together with the figures. Unfortunately I forgot to highlight the caption changes in blue... can you just completely replace the old captions with the new ones? We worked hard to make the captions as short as possible, while retaining their accuracy.

...... Also if you want the new captions/figures as Word rather than PDF, please say (the Word file is large and very slow to open on my PC).

While the contemporaneous Climategate emails show a team of scientists struggling to put Chapter 6 of AR4 together, in August several discussions on Climate Audit started to raise the issues of Jones' dataset in a thread on Warwick Hughes work on sea temperatures, a thread on hurricanes and temperature measurements of the troposphere. Commenter Tim Ball chimes in to defend Warwick and remind readers of Jones' refusal of Warwick's request for data:

Tim Ball

Posted Aug 12, 2006 at 6:40 PM | Permalink

Would Steve Bloom and the Sierra Club please work to get Mann to release the codes and Phil Jones to disclose how he calculated the increase in global mean temperature instead of besmirching other people. And while they're at it, they can help the research by providing funding for an audit of the climate models. Oh and by the way don't respond to this with an ad hominem attack.

Climate Audit regular Willis Eschenbach makes a series of comments throughout the month targeting Jones' data, the lack of its availability and the importance of the UHI problem. Eschenbach is about to get involved directly.

At McIntyre's blog a discussion of Steve's request for data to the NAS panel and their refusal to even ask for the data was posted in mid September. And John A, a CA regular, made the suggestion that perhaps Steve should move on from informal requests and make a formal FOIA request for the information he sought.

John A

Posted Sep 18, 2006 at 1:07 PM

> **What I'd like to know is when will Steve actually make an FOIA request for all of this data and break this particular logjam?**

McIntyre doesn't—yet, but frustrated by Jones's refusal to provide data to Warwick Hughes and perhaps inspired by John A's comment, Eschenbach sent the following FOIA request to Jones.

> *I would like to obtain a list of the meteorological stations used in the preparation of the HadCRUT3 global temperature average, and the raw data for those stations. I cannot find it anywhere on the web. The lead author for the temperature average is Dr. Phil Jones of the Climate Research Unit.*
>
> *Many thanks, Willis Eschenbach*

Eschenbach will not reveal this request for five months, waiting until his request is refused before telling anyone he had even made a request.

2007: The Year of UHI and FOIA

There is no formal agenda in the Climate Wars. Scientists and bloggers both choose and pursue topics according to their interests, and those looking for a grand schema that will guarantee a certain amount of time and attention for each topic or phase will be disappointed. Those following discussion of climate issues will easily understand that there are so many different topics being debated at any one time that some disappear off the radar screen for weeks or even months at a time while other topics grab the attention of those involved. The issue of Jones's refusal to release the underlying data stayed somewhat dormant until early 2007. But the topic of UHI was in the forefront of mails between Susan Solomon of NOAA and Jones. Solomon and others were preparing for a presentation in Paris and the issue of UHI came up. How was it dealt with in studies? During the course of the mails Solomon gets the size of the effect wrong by an order of magnitude, but we can see the centrality of Jones 1990 with regard to this important subject. A lot stands on Jones' 4 page paper in Nature.

> *To: Susan Solomon <Susan.Solomon@noaa.gov>*
>
> *Subject: Re: Science presentation for Paris*
>
> *Date: Mon, 08 Jan 2007 15:31:18 -0700*
>
> > *One too many 0's. 0.005.*
>
> *Phil,*
>
> > *Thanks. This comes up both in the presentation and in SPM language.   A suggested merge of Phil's text below with the SPM language we have implies replacing the sentence on page SPM-5, 6-7 with the following proposal: Sites affected by the urban heat island effect are identified and excluded from these averages, so that remaining uncertainties due to this effect are negligible (less than 0.0005∞C per decade). This would address several comments asking us to explain what is done with UHI.   OK?*

Susan

Jones clears up the issue with the following:

> *At 3:52 PM +0000 1/8/07, Phil Jones wrote:*
>
> *Kevin, Susan,*
>
> *On the UHI (slide 9) we should probably change the middle bullet. ... Middle bullet currently says*
>
> *o Major influences are identified and excluded from the records used to create the continental and global values Perhaps we should refer directly to David Parker's paper on UHIs, where he couldn't detect any difference in trends (averaged for 200+ cities) in temperatures on calm nights (when you'd expect the biggest effect) compared to windy nights (when you'd expect the least). There are two aspects to the major influences. 1. Some sites are removed. This isn't many as a % of the total (about 1%). 2. We include in Brohan et al (2006) an estimate of urbanization in the calculation of the errors. This is 0.0055 deg C/decade since 1900. It is a one-sided 'error'. If you look very closely the error range in this paper and in some of the Ch 3 figures is slightly one-sided. This figure comes from Jones et al. (2001), which came from Jones et al. (1990). Difficulty with all UHI work is that there are countless papers looking at individual sites - which generally use a site in the city centre. This site is rarely one used in the dataset - generally an airport is instead. It is made worse by then looking at individual days and not monthly averages. Only Jones et al. (1990), Parker (2005,2006) and Peterson have looked at large scales. So Affected site are identified and excluded from the records used to create the continental and global values (as not all sites are tested, part of the error range assumes an urban component of 0.0055 deg C/decade)*
>
> *Cheers   Phil*

In the January time period there was also a minor side issue at play on the web with regards to a hand drawn diagram made by Hubert Lamb that had found its way into one of the early IPCC reports. The diagram shows a Medieval Warm Period (MWP) that is higher than the current temperatures. We note this not because it shows a higher MWP, although that's certainly of interest to the discussion. It is directly relevant because it shows that the IPCC is willing to rely on non peer reviewed literature when it's published by Lamb, the first director of CRU. This sheds some light on the psychology of those at CRU. Tom Wigley, a former CRU employee, explains to Jones the history behind this diagram and how it was wrong. It illustrates that the organization is concerned more with image than with science. That they loathe admitting mistakes, even minor ones, and when the science absolutely demands it, they find means to soften the blows. Reputation takes precedence over science. Where the junk science of tobacco companies was produced for profit, inside CRU reputation has the same corrosive effect. The diagram in question was wrong. Scientists at CRU knew it was wrong, but they did not want to offend Lamb by correcting him by publishing the correction in a mainstream journal.

> *Phil, I see the problems with this[Lamb Diagram] in terms of history, IPCC image, skeptix, etc. I'm sure you can handle it. In doing so, you might consider (or not) some of these points. (1) I think Chris Folland is to blame for this. The issue is not our collective ignorance of paleoclimate in 1989/90, but Chris's ignorance. The text that was in the 1990 report (thanks for reminding us of this, Caspar) ameliorates the problem considerably. (2) Nevertheless, 'we' (IPCC) could have done better even then. The Rothlisberger data were available then -- and could/should have been used. (3) We also already knew that the Lamb UK record was flawed. We published a revision of this -- but never in a mainstream journal because we did not want to offend Hubert. I don't have the paper to hand, but I think it is ...*

At Climate Audit, UHI was also on the agenda for the spring and McIntyre turned out a large number of posts on the topic. Targeting Jones's 1990 work, McIntyre had posted on the statistics of record breaking temperatures and Eschenbach was pounding on the issue of data access and UHI.

A rather technical post from Climate Audit regarding the adjustments that must be made to temperature records in order to calculate global averages (such as time of day that temperatures are taken, changes of or the collection station, and corrections for urban heat island effects where appropriate):

Willis Eschenbach

Posted Jan 17, 2007 at 3:31 AM | Permalink | Reply

….. Steve has shown these types of errors both in the past and today, and since the weather observers are human, and since the people typing up the records hit the wrong key occasionally, and since weather observation in many parts of the globe are not well trained and sometimes are drunk or just don't care, and since records from the last century were written by hand and are sometimes illegible, and since Jones refuses to let outsiders see his data (why…….

…….. I thought, hey, maybe that's just me, which is why I asked the question.

Unfortunately, as noted above, Phil Jones and his merry men won't let outsiders see the dataset, so we have no way to know how corrupted the underlying data is, or what kind of quality control procedures they have in place. For me, the very first thing I do with a dataset, before taking any other measures to identify possibly bad data, is to graph it for a preliminary scan … but as Steve has shown, they haven't even done that.

w.

Willis Eschenbach

Posted Jan 21, 2007 at 10:48 PM

Well, I've been thinking more about the HadCRUT3 dataset and the errors therein. I decided to take a look at the coverage, to see what effect that might have on the data. …..

The HadCRUT3 coverage has decreased, …….

...... Coverage is minimal in the north, and almost nonexistent in the south. In those regions, it's almost all computer model results, not data.

4) Since Jones refuses to reveal his data, we have no chance of a replication of his underlying methods.

5) Since the information in the error analysis paper is inadequate to determine the exact methods used for the calculation of the averages and errors, there is no way to determine if they have been done correctly or reported accurately. There may be a more detailed description of the averaging method elsewhere, but I have been unable to locate it.

w.

Willis Eschenbach

Posted Jan 22, 2007 at 7:51 PM | Permalink | Reply

Steve S, you say:

RE: #87 – There is a story within the story. What we see here is the truth about "urban" vs "rural" stations. The truth is, "rural" stations are really stations located in small to medium sized towns, for the most part. As there are few towns of any sort north of 70, well, there you have it. Almost no stations.

As always in the climate world, there is more than one story within the story.

One is that the northernmost US station at Barrow, Alaska, is known to be badly affected by UHI, and thus corrupts the record much, much more than it would if there were more stations around.

The second is that UHI occurs more easily when the weather is cold. Adding a bunch of houses that are kept at 70°F (20°C) won't make much difference if that is the ambient temperature, but will have a big effect if the temperature is -40° (either C or F, take your pick, they're the same ...)

The third, and most important, is how Jones et al. treat the UHI error. He gives it a value of 0°C up to the year 1900, and increasing linearly to ±0.06° in 2006. This seems strange for a couple of reasons:

1) He provides no justification other than his own 1990 study for the purported size of the error.

2) He treats the error as symmetrical:

The urbanization uncertainty could be regarded as one sided: stations cannot be "too rural" but may inadvertently be "too urban" (Jones et al., 1990; Peterson et al., 1999). However, because some cold biases are also possible in adjusted semi-urban data, we conservatively model this uncertainty as symmetrical about the optimum average. We assume that the global average LAT uncertainty (2 sigma) owing to urbanization linearly increases from zero in 1900 to 0.1°C in 1990 (Jones et al, 1990), a value we extrapolate to 0.12°C in 2000 (Figure 1a).

While stations cannot be "too rural", it is quite possible that the majority of "rural" station need adjustment for heat island effects. Also, a symmetrical error assumes that he has done the UHI adjustment in some basically correct manner. But we can't tell, because ...

3) As far as I know he has never provided a detailed description of exactly how he is doing the adjustment for the UHI.

w.

Host McIntyre weighs in, recounting his prior communication with Jones prior to the publication of MM03. And he notes a connection between Jones' surface temperature work and climate reconstructions that very few appreciate.

Steve McIntyre

Posted Jan 24, 2007 at 7:53 AM

.... A couple of years ago, before I had any notoriety, I asked Phil Jones for the data used in his UHI study. He said that it was on a diskette somewhere but it would be too hard to find. At the time I wasn't aware of the systematic obstruction and didn't pursue the matter.

Given that Jones' work has been funded by the U.S. Department of Energy (starting with the Carbon Dioxide Information Center of the Oak Ridge Nuclear Lab in Tennessee), the flaccid administration of the DOE is equally or more to blame for acquiescing in Jones' general intransigence.

An interesting sidenote on HadCRU. Briffa's reported correlation between gridcell temperature and foxtail chronology did not hold for the HadCRU data said to have been used in Osborn and Briffa. When I pursued this, it turned out that he used CRUTem instead of HadCRU data. The CRUTem for his gridcell started only in 1888 while the HadCRU started in 1870. He said that the HadCRU dataset had some bad data for 1870-1888. I tried to get an explanation of what was wrong with HadCRU and right with CRUTem but got nowhere with Science. Neither Briffa nor Science felt that a corrigendum was required to provide a correct identification of the data actually used in his calculation.

When scientists such as Mann and Briffa try to reconstruct past temperatures they rely on the instrumented series-- the historical record. As McIntyre points out sometimes they use different versions of the instrument record. The concern is this: a bad temperature series will compound errors in a reconstruction. Simply, reconstructions of past temperature depend upon correlations with accurate historical instrument records. If the historical record is corrupt then the reconstruction will be corrupt. So looking at Jones's temperature data is not only important to understand if the problem of UHI is handled correctly, it's important because climate reconstructions depend on its accuracy. Continuing, the Climate Audit faithful chimed in:

Nordic Posted Jan 24, 2007 at 11:22 AM | Permalink |

Thanks for the link to the temp. datasets. I was curious to see what stations they used in my area, but was not able to locate a map or listing. Do you know where I can find such a list? I am most interested in HasCRUT3

Earle Williams Posted Jan 24, 2007 at 12:41 PM

Nordic, I'm a babe in the woods with respect to these datasets. I haven't even figured out how to read the HadCRUT3 gridcell data into R. I suggest rooting around the HadCRU web site and emailing them if the info isn't available.

Luck, Earle

And McIntyre responds to his readers citing the famous reply that Jones gave Warwick Hughes Feb 21, 2005:

Steve McIntyre Posted Jan 24, 2007 at 1:46 PM

You will have no luck trying to get station identifications from HadCRU. This has been an ongoing complaint. Jones said when asked: We have 25 years invested in this – why should we let you see the data when your only objective is to find something wrong with it"?

Nordic Posted Jan 24, 2007 at 3:17 PM | Permalink | Reply Steve: RE #123 I have seen that quote posted here before, but had no idea they didn't even release the names of the stations they had used. That is just incredible. Not releasing the details of how they have adjusted and corrected data is one thing, but not even allowing people to see what stations were used and experiment with different assumptions themselves is preposterous.

Earle Williams Posted Jan 24, 2007 at 3:30 PM | Permalink | Reply Re #125 Nordic, I agree with you 100%, but it is at least worth a shot. You may be lucky enough to get a clerk or student to send you the info.

Earle

As luck would have it there was a "lurker" on the site that day. An anonymous person, here named "bez" who reads the site but rarely comments. He tells the readers about the tool they will need to get the data freed: he points out that CRU (the Climate Research Unit) is a part of UEA (the University of East Anglia) and hence subject to FOIA (the UK's Freedom of Information Act) requests from anyone in the world:

bez

Posted Jan 25, 2007 at 2:43 AM

Hi All, been lurking for a while now.

Re: 125 and previous posts

CRU is part of UEA hence

http://www1.uea.ac.uk/cm/home/services/units/is/strategies/infregs/foi/

May or may not be some use for getting data out of them.

Eschenbach, who wrote an FOIA request 5 months earlier, already aware of the FOIA requirements on UEA, reveals his September 28 request in a post on Feb 18, 2007.

Willis Eschenbach

Posted Feb 18, 2007 at 2:57 AM | Permalink | Reply

Well, I'm on holiday in Fiji, followed by two weeks work in Solomon Islands, so I'll be in and out. Just wanted to note that I filed a Freedom of Information request for Jones et al.'s HadCRUT data. My request said:

I would like to obtain a list of the meteorological stations used in the preparation of the HadCRUT3 global temperature average, and the raw data for those stations. I cannot find it anywhere on the web. The lead author for the temperature average is Dr. Phil Jones of the Climate Research Unit.

Many thanks,

w.

Eschenbach continues and copies his reply from CRU/UEA, dated Feb 10th 2007.

Dear Mr. Eschenbach

FREEDOM OF INFORMATION ACT 2000 – INFORMATION REQUEST (FOI_07-04)

Your request for information received on 28 September has now been considered and I can report that the information requested is available on non-UEA websites as detailed below.

The Global Historical Climatology Network (GHCN-Monthly) page within US National Climate Data Centre website provides one of the two US versions of the global dataset and includes raw station data. This site is at. http://www.ncdc.noaa.gov/oa/climate/ghcn-monthly/index.php

This page is where you can get one of the two US versions of the global dataset, and it appears that the raw station data can be obtained from this site.

Datasets named ds564.0 and ds570.0 can be found at The Climate & Global Dynamics Division (CGD) page of the Earth and Sun Systems Laboratory (ESSL) at the National Center for Atmospheric Research (NCAR) site at: http://www.cgd.ucar.edu/cas/tn404/

Between them, these two datasets have the data which the UEA Climate Research Unit (CRU) uses to derive the HadCRUT3 analysis. The latter, NCAR site holds the raw station data (including temperature, but other variables as well). The GHCN would give their set of station data (with adjustments for all the numerous problems).

They both have a lot more data than the CRU have (in simple station number counts), but the extra are almost entirely within the USA. We have sent all our data to GHCN, so they do, in fact, possess all our data.

In accordance with S. 17 of the Freedom of Information Act 2000 this letter acts as a Refusal Notice, and the reasons for exemption are as stated below

Exemption Reason

s. 21, Information accessible to applicant via other means. Some information is publicly available on external websites

If you have a complaint about the handling of your enquiry then please contact me at: University of East Anglia Norwich NR4 7TJ Telephone 0160 393 523 E-mail foi@uea.ac.uk

You also have a right of appeal to the Information Commissioner at: Information Commissioner's Office Wycliffe House Water Lane Wilmslow Cheshire SK9 5AF Telephone: 01625 545 700 http://www.ico.gov.uk

Yours sincerely

David Palmer Information Policy Officer University of East Anglia

Since the leak of the Climategate emails, many defenders of The Team have said that the data skeptics have asked for in order to replicate their work has always been available for inspection. Indeed, a lot of data is available on various government websites. But often, as in this case, what skeptics need is actually simpler—they need to know what data was actually used in calculations. They do know where the data lives—but not what data scientists choose to use or exclude in their studies.

As Eschenbach will point out there are several problems with this response. First and foremost Eschenbach requested a list of **the stations** and the raw data. A station is merely a name or a number associated with a physical location in the world where the data is collected. The raw data is the actual temperature collected at this station. CRU made several strange statements. They point Eschenbach to a US source (GHCN and another at NCAR) and simultaneously claim that they both get their data from GHCN and supply data to GHCN. Further they note that GHCN has more stations than CRU. CRU is a subset of GHCN. Eschenbach and others have known this and what they are requesting is the **list of stations** that CRU actually use. In effect, GHCN may have raw data for 6,000 stations, but CRU will use far fewer. Eschenbach wanted to know exactly which ones CRU used. They pointed him to the vast group they had selected from and said, in effect, we got them from that large group. But which ones? Jones and the FOIA officer are employing the following dodge. Jones used a subset of data which he claimed to get from the thousands of stations at GHCN. When Eschenbach asked for the list of which stations and a copy of the raw data for those stations, CRU responded by pointing at the collection of 6000 stations. They refused to say which stations they used. They just pointed at a stack of 1000s and said that the sites used by Jones were somewhere in that vast collection.

While this is playing out on Climate Audit, Jones and Peterson are reading CA. Watching the commenters struggling to get access to the data, Peterson writes to Jones the following mail and attaches a cartoon that lampoons various figures in the climate change debate that The Team considered 'bad guys', unaware of the problems that loom ahead of them in the fall of 2009. They were too caught up in the process of AR4 to realize the problem these data requests posed for them. In the end, ironically, Jones would be marooned as colleagues such as Michael Mann would throw him under the bus for suggesting that data be deleted rather than shared and for later going so far as suggesting that emails be deleted:

> *From: "thomas.c.peterson" <Thomas.C.Peterson@noaa.gov>*
>
> *To: Phil Jones <p.jones@uea.ac.uk>   Subject: [Fwd: Marooned?]*
>
> *Date: Mon, 19 Feb 2007 11:10:02 -0500*
>
> *Hi, Phil, I thought you might enjoy the forwarded picture and related   commentary below. read some of the USHCN/GISS/CRU brouhaha on web site you sent us.   ....*
>
> *Tom*

Figure 5: Photoshopped Representation of Team Opponents

Infuriated by CRU's FOIA refusal letter, Eschenbach sends another letter, dated Feb 10th 2007, the same day he received the rejection

Dear Mr. Palmer:

Thank you for your reply (attached below). However, I fear that it is totally unresponsive. I had asked for a list of the sites actually used. While it may (or may not) be true that "it appears that the raw station data can be obtained from [GHCN]", this is meaningless without an actual list of the sites that Dr. Jones and his team used.

The debate about changes in the climate is quite important. Dr. Jones' work is one of the most frequently cited statistics in the field. Dr. Jones has refused to provide a list of the sites used for his work, and as such, it cannot be replicated. Replication is central to science. I find Dr. Jones attitude quite difficult to understand, and I find your refusal to provide the data requested quite baffling.

You are making the rather curious claim that because the data "appears" to be out on the web somewhere, there is no need for Dr. Jones to reveal which stations were actually used. The claim is even more baffling since you say that the original data used by CRU is available at the GHCN web site, and then follow that with the statement that some of the GHCN data originally came from CRU. Which is the case? Did CRU get the data from GHCN, or did GHCN get the data from CRU?

Rather than immediately appealing this ruling (with the consequent negative publicity that would inevitably accrue to CRU from such an action), I am again requesting that you provide:

1) A list of the actual sites used by Dr. Jones in the preparation of the HadCRUT3 dataset, and

2) A clear indication of where the data for each site is available. This is quite important, as there are significant differences between the versions of each site's data at e.g. GHCN and NCAR.

I find it somewhat disquieting that an FOI request is necessary to force a scientist to reveal the data used in his publicly funded research ... is this truly the standard that the CRU is promulgating?

Thank you for your cooperation in this matter.

Willis Eschenbach

Eschenbach is asking for the fundamental data that one needs to check and verify one of the most important pieces of data in climate science: the data backing up the claim that the globe has gotten warmer since 1850. CRU refused his request, saying that the data was already available and they seem to have deliberately misunderstood his exact request. They withheld the names of the exact stations CRU used, and merely said they were in the massive databases at GHCN and NCAR. Eschenbach gets his second reply:

In regards to the "gridded network" stations, I have been informed that the Climate Research Unit's (CRU) monthly mean surface temperature dataset has been constructed principally from data available on the two websites identified in my letter of 12 March 2007. Our estimate is that more than 98% of the CRU data are on these sites.

The remaining 2% of data that is not in the websites consists of data CRU has collected from National Met Services (NMSs) in many countries of the world. In gaining access to these NMS data, we have signed agreements with many NMSs not to pass on the raw station data, but the NMSs concerned are happy for us to use the data in our gridding, and these station data are included in our gridded products, which are available from the CRU web site. These NMS-supplied data may only form a very small percentage of the database, but we have to respect their wishes and therefore this information would be exempt from disclosure under FOIA pursuant to s.41. The World Meteorological Organization has a list of all NMSs.

CRU now is changing its story. At first they argued that all their data was at GHCN or NCAR. In fact CRU assembled the data under DOE contract. But now, they are arguing that the data cannot be released because some of it, 2%, was procured under confidentiality agreements. But which countries had supplied data under these agreements? And why should CRU withhold all the data if only 2% were covered by agreements? A redacted release would cover both their obligations under FOIA and whatever agreements they had. In the extreme, for example, according to CRU logic, if one station out of several thousand was covered by an agreement, they would be justified in withholding all the open data. Finally their last sentence suggests that Eschenbach should look to National Meterological Services (NMS) organizations to get the data. But without the names of countries and stations which CRU used, Eschenbach cannot know who to ask. Eschenbach replied:

> While it is good to know that the data is available at those two web sites, that information is useless without a list of stations used by Jones et al. to prepare the HadCRUT3 dataset. As I said in my request, I am asking for:
>
> 1) A list of the actual sites used by Dr. Jones in the preparation of the HadCRUT3 dataset, and
>
> 2) A clear indication of where the data for each site is available. This is quite important, as there are significant differences between the versions of each site's data at e.g. GHCN and NCAR.
>
> Without knowing the name and WMO number of each site and the location of the source data (NCAR, GHCN, or National Met Service), it is not possible to access the information. Thus, Exemption 21 does not apply – I still cannot access the data.
>
> I don't understand why this is so hard. All I am asking for is a simple list of the sites and where each site's data is located. Pointing at two huge piles of data and saying, in effect, "The data is in there somewhere" does not help at all.

To clarify what I am requesting, I am only asking for a list of the stations used in HadCRUT3, a list that would look like this:

WMO# Name Source 58457 HangZhou NCAR 58659 WenZhou NCAR 59316 ShanTou GHCN 57516 ChongQing NMS

etc. for all of the stations used to prepare the HadCRUT3 temperature data.

That is the information requested, and it is not available "on non-UEA websites", or anywhere else that I have been able to find.

I appreciate all of your assistance in this matter, and I trust we can get it resolved satisfactorily.

Best regards,

That letter as well did not move the case forward so Eschenbach sent another:

Dear Mr. Palmer:

It appears we have gone full circle here, and ended up back where we started.

I had originally asked for the raw station data used to produce the HadCRUT3 dataset to be posted up on the UEA website, or made available in some other form.

You refused, saying that the information was available elsewhere on non-UEA websites, which is a valid reason for FOI refusals.

I can report that the information requested is not available on non-UEA websites as detailed below.

Your most recent letter (Further _information_ letter_final_ 070418_rev01. doc), however, says that you are unable to identify the locations of the requested information. Thus, the original reason for refusing to provide station data for HadCRUT3 was invalid.

Therefore, since the information requested is not available on non-UEA websites, I wish to re-instate my original request, that the information itself be made available on your website or in some other form. I understand that a small amount of this data (about 2%, according to your letter) is not available due to privacy requests from the countries involved. In that case, a listing of which stations this applies to will suffice.

The HadCRUT3 dataset is one of the fundamental datasets in the current climate discussion. As such, it is vitally important that it can be peer reviewed and examined to verify its accuracy. The only way this can be done is for the data to be made available to other researchers in the field.

Once again, thank you for your assistance in all of this. It is truly not a difficult request, and is fully in line with both standard scientific practice and your "CODE OF PRACTICE FOR RESPONDING TO REQUESTS FOR INFORMATION UNDER THE FREEDOM OF INFORMATION ACT 2000". I am sure that we can bring this to a satisfactory resolution without involving appeals or unfavorable publicity.

My best regards to you, w.

David Palmer responds and in the end Eschenbach gets half of his wish granted. He gets the list of stations, but not the data: In the end the station list, replete with errors, will be posted on the web, but not until Oct 1, 2007 almost 6 months after agreeing to do so. Continuing with the responses:

Dear Mr. Eschenbach FREEDOM OF INFORMATION ACT 2000 – INFORMATION REQUEST (FOI_07-04)

Further to your email of 14 April 2007 in which you re-stated your request to see

"a list of stations used by Jones et al. to prepare the HadCRUT3 dataset" I am asking for: 1) A list of the actual sites used by Dr. Jones in the preparation of the HadCRUT3 dataset, and 2) A clear indication of where the data for each site is available. This is quite important, as there are significant differences between the versions of each site's data at e.g. GHCN and NCAR."

In your note you also requested "the name and WMO number of each site and the location of the source data (NCAR, GHCN, or National Met Service)",

I have contacted Dr. Jones and can update you on our efforts to resolve this matter.

We cannot produce a simple list with this format and with the information you described in your note of 14 April. Firstly, we do not have a list consisting solely of the sites we currently use. Our list is larger, as it includes data not used due to incomplete reference periods, for example. Additionally, even if we were able to create such a list we would not be able to link the sites with sources of data. The station database has evolved over time and the Climate Research Unit was not able to keep multiple versions of it as stations were added, amended and deleted. This was a consequence of a lack of data storage in the 1980s and early 1990s compared to what we have at our disposal currently. It is also likely that quite a few stations consist of a mixture of sources.

I have also been informed that, as the GHCN and NCAR are merely databases, the ultimate source of all data is the respective NMS in the country where the station is located. Even GHCN and NCAR can't say with precision where they got their data from as the data comes not only from each NMS, but also comes from scientists in each reporting country.

In short, we simply don't have what you are requesting. The only true source would be the NMS for each reporting country. We can, however, send a list of all stations used, but without sources. This would include locations, names and lengths of record, although the latter are no guide as to the completeness of the series.

This is, in effect, our final attempt to resolve this matter informally. If this response is not to your satisfaction, I will initiate the second stage of our internal complaint process and will advise you of progress and outcome as appropriate. For your information, the complaint process is within our Code of Practice and can be found at: http://www1. uea.ac.uk/ polopoly_ fs/1.2750! uea_manual_ draft_04b. pdf

Yours sincerely David Palmer Information Policy Officer University of East Anglia

The FOIA correspondence with Eschenbach is important because it illustrates the very point made by McIntyre and other commenters at his site. The science behind calculating a global average is trivial. It is more properly an accounting task. But Jones and CRU appear to have made a total hash of the record keeping. If their responses to FOIA requests are to be believed, they don't have a list of the sites they currently use. They don't have a version control system that allows them to reconstruct what they have done over the years. And they can't tell the ultimate source of their data. Their work cannot be double checked. It must be taken on trust. And finally, they don't believe that GHCN and NCAR can say with precision where their data came from. In summary, the IPCC and the huge bulk of papers on climate science depend upon CRU's version of the global temperature. That data series uses data from a variety of sources, most notably GHCN. And according to the FOIA officer of CRU, CRU hasn't kept adequate records to reconstruct its work, and GHCN cannot state with certainty where they got their data from. The global temperature index is unaccountable. The NCDC houses GHCN and reports to NOAA. Phil Jones is a member of NOAA's standing advisory board on data archiving.

Sadly, we tend to believe Jones and CRU, and this is in many ways more troubling than if they had done something intentionally wrong. If a crime or unethical behavior had occurred, then part of the resolution of a case would involve putting things right in the data. But what Jones and CRU lay out for us is an unfixable dilemma—unless the entire process is started from scratch. And we tend to also think that their realization of the sorry state of their data is one of the real reasons for their reluctance to make their data available.

Eschenbach appealed the FOIA decision and his appeal was denied. A few short days after Eschenbach announced in Feb of 2007 that he had sent CRU an FOIA request for all the stations they used in the global index, McIntyre announces his own FOIA action. Taking a hint from the runaround Eschenbach received, McIntyre delivers a very targeted request. He requests the stations and the data for Jones's seminal 1990 paper—the paper that Warwick Hughes had been criticizing, the paper that asserted that UHI did not contaminate the global record, much: On March 9, McIntyre announces his Feb 22 letter:

On Feb. 22, 2007, I sent the following request to Phil Jones:

Dear Phil, a couple of years ago, I requested the identities and data for the Russian, Chinese and Australian networks studied in Jones et al Nature 1990 on urbanization. At the time, you said that it would be unduly burdensome to locate the information among your diskettes as the study was then somewhat stale. However, I notice that Jones et al 1990 has been cited in IPCC AR4 (in the section where you were a Coordinating Lead Author) and continues to be cited in the literature (e.g. Peterson 2003).

Accordingly, I re-iterate my request for the identification of the stations and the data used for the following three Jones et al 1990 networks:

1. the west Russian network 2. the Chinese network 3. the Australian network

For each network, if a subset of the data of the data was used, e.g. 80 stations selected from a larger dataset, I would appreciate all the data in the network, including the data that was not selected.

In each case, please also provide the identification and data for the stations used in the gridded network which was used as a comparandum in this study.

Thank you for your attention.

Steve McIntyre

CRU responded by transferring authority to a different set of regulations, known as the EIR regulations. CRU had 20 more days to respond. McIntyre launches a series of posts to compare the results of Jones 1990 with the data available from GHCN. Jones, cited in IPCC literature, was a foundational study. Cited by Peterson 2003, also an IPCC source:

> Jones et al. (1990) determined that the impact of urbanization on hemispheric temperature time series was, at most, 0.058 deg C century-1. This result was based on the work of Karl et al. (1988) for the United States and further analysis of three other regions: European parts of the Soviet Union, eastern Australia, and eastern China. In none of these three regions was there any indication of significant urban influence in either of the two gridded time series relative to the rural series' (Jones et al. 1990). The homogeneity assessments varied with regions. The data for one region were assessed for artifacts due to factors such as site moves or changing methods used to calculate monthly mean temperatures.' Another region used data from stations with few, if any, changes in instrumentation, location or observation times.' The homogeneity of data used in the third region was not discussed. Their results showed that the urbanization influence is, at most, an order of magnitude less than the warming seen on a century scale

While the bureaucratic wheels ground on at CRU, McIntyre put out a series of posts on Russian sites and Chinese sites, all of which raise serious questions about Jones' 1990 paper. CRU responded with a rejection letter that follows the same pattern as those received by Eschenbach.

Dear Mr. McIntyre

ENVIRONMENTAL INFORMATION REGULATIONS 2004 – INFORMATION REQUEST (FOI_07-09; EIR_07-02)

Your request for information has now been considered and the information requested is enclosed.

Some of the information requested cannot, however, be disclosed and, Pursuant to Regulation 12, Environmental Information Regulations 2004, I am not obliged to supply this information. The exemptions are clearly indicated within the attached document and the reasons for exemption are as stated below

Exemption Reason

Reg. 6(1)b: Information already publicly available & easily accessible to the applicant

Reg. 12(4)a: Information not held by the authority

The reason for claiming Regulation 6(1)(b) is that the station specific raw (i.e. daily) urban' data requested is already accessible on publicly available websites, specifically: 1) The Global Historical Climatology Network (GHCN-Monthly) page within US National Climate Data Centre website at: http://www.ncdc.noaa.gov/oa/climate/ghcn-monthly/index.php, and, 2) the Climate & Global Dynamics Division (CGD) page of the Earth and Sun Systems Laboratory (ESSL) at the National Center for Atmospheric Research (NCAR) site at: http://www.cgd.ucar.edu/cas/tn404/

In regards Regulation 12(4)(a), the information from rural data stations no longer exists in the form requested at the University of East Anglia.

The public interest in claiming these exemptions is clear; in the case of Reg. 6(1)(b), information can be provided to the requester faster, and without diverting resources of the University than if the University were to provide this information directly. Clearly, we cannot provide information we do not possess, and the public interest is not at issue.

I apologise that not all of your request will be met but if you have any further information needs in the future then please contact me.

Yours sincerely

David Palmer Information Policy Officer University of East Anglia

As McIntyre points out, their first sentence makes no sense. They did not enclose any information with their refusal. They denied the request, using the same logic they tried with Eschenbach. McIntyre was requesting **the names** of the few hundred stations used in Jones paper. And Palmer points him to a collection of thousands of possible stations: It is as if McIntyre has asked a librarian for a book checked out by Jones and the librarian responds by shrugging and pointing to the library: He writes Palmer:

Dear Mr Palmer,

Your reply is unresponsive to my request and not in accordance with your policies. I hereby request that you re-consider your refusal to provide the following information:

A) the identification of the stations ... for the following three Jones et al 1990 networks:

1. the west Russian network 2. the Chinese network 3. the Australian network

B) identification ... of the stations used in the gridded network which was used as a comparandum in this study

While the data for these stations may or may not exist at GHCN and/or CGD (and such data may or may not be in the version used in Jones et al 1990), it is impossible to identify the stations used in Jones et al 1990 by inspection of the GHCN or CGD data sets and your application of Reg 6(1)b cannot be justified. I would be surprised if Jones et al no longer even have a record of what stations were used in their study. Acccordingly, I appeal your ruling.

Yours truly,

Stephen McIntyre

McIntyre goes one better and files a complaint with Nature, the science journal where Jones' paper was published. Jones' 1990 as published in Nature is in violation of Nature's materials policy, which requires authors to make data available. Whether the request to Nature had anything to do with changing CRU's opinion on McIntyre's request is unclear. Had they forced Jones to reply they would have set a precedent. In the end, CRU relented and claimed they found the sites (Russian, Chinese and Australian) and promised to post them on the web by April 13th.

FREEDOM OF INFORMATION ACT 2000 – INFORMATION REQUEST (FOI_07-09)

In your email of 12 March 2007, you stated that you wished to see "A) the identification of the stations … for the following three Jones et al 1990 networks: 1. the west Russian network 2. the Chinese network 3. the Australian network B) identification … of the stations used in the gridded network which was used as a comparandum in this study"

As part of our obligation to, at first instance, try to resolve appeals informally, I have been in discussion with the relevant members of UEA and, upon further investigation and work, we have uncovered the annual input data for the paper of Dr. Jones from 1990.

This information includes locations of the sites and the annual temperature values for those sites. We are putting this information on the UEA website with accompanying explanatory text and anticipate that, due to the Easter holidays, this will be available to the public no later than Friday 13 April. We are, of course, happy to supply the url for this information as soon as we have it.

We do not have any information about why the sites for the 1990 paper were selected as Dr. Jones is unaware of how his collaborators selected the sites.

We hope that that this information satisfies your original request of 22 February that it will resolve this matter amicably and to your satisfaction.

Yours sincerely

David Palmer

Information Policy Officer

University of East Anglia

So, over two years after denying the data to Warwick Hughes on the grounds that Hughes' motives were suspect, that he wanted to find mistakes in Jones' work, CRU relents and agrees to post the data that forms the basis of Jones' 1990 paper. The other critical thing to recognize here is the record keeping problem that CRU and Jones seem to have. Data once thought lost is now found. The data, however, is only annual data and Jones is 'unaware' of how his collaborators selected the sites. Yet Jones 1990 is supposed to be a study that investigates whether or not site selection matters. As it stands in April of 2007 one of the outstanding issues appears to be fully resolved: CRU will deliver the station names and the annual data for Jones' 1990 study. On the larger request for all of CRU's data, one open issue remains: CRU has agreed to publish a list of stations for the entire globe, but they refuse to release the actual station data. That refusal drives the rest of the story.

After years of asking scientists directly for their data and being denied on utterly personal grounds, after denials from journals who refused to follow their own data release policies, after denials by panels of experts impaneled by the US congress to investigate the question of replicating the science, Steve McIntyre and his readers found a tool that appeared to work: FOIA. The success McIntyre and his team had with FOIA with regards to the UHI question will also fuel a new round of FOIAs targeting US agencies and the IPCC process itself in the years to come

For those who find this all somewhat surprising, we note that in some of the Climategate emails, Phil Jones writes his colleagues that, 'after one or two half-hour sessions' he was able to convince the Freedom of Information officer that skeptics such as those found at Climate Audit did not need to be treated under the law, a somewhat dubious conclusion. We also see elsewhere that the FOIA officer, David Palmer, will use leading questions to insure that more information is not included in FOIA responses, again, quite disingenuously. And for those who care to speculate about motives, the 'deny and delay' principle does serve to push McIntyre's criticism of The Team's work past the publication date of AR4. As McIntyre's previous criticism of Michael Mann's Hockey Stick had proven devastating to their work, perhaps it was better to get it published before McIntyre had a chance to find holes in it.

CHAPTER SIX: AN ARMY OF DAVIDS

Cheat Sheet: McIntyre got gamed as reviewer of the IPCC AR4 report, with The Team working around him to get what they needed into the influential report. McIntyre had lost a battle, and the activist position on global warming surged to its highest level of support, based on the IPCC's assurances that even if there were minor problems with the Hockey Stick chart, other reconstructions of past temperatures supported its central conclusion—that temperatures today were unprecedented, and were going to get worse. McIntyre does not appear to be discouraged by this, continuing to use something similar to forensic accounting procedures, and begins to use the new tool, FOIA requests, to continue his search for the data. Ironically the tricks they use in the review process only serve to inspire McIntyre to expose them, and the communication that occurs outside the process is so vital that Jones will counsel people to delete these mails. It's always the coverup. And McIntyre's starting to get help—not just from regulars at Climate Audit, but larger numbers of sympathetic bystanders who are regulars at Watt's Up With That, the most popular climate weblog in town.

In Chapter 6 we introduce the Army of Davids that will start the laborious process of documenting all the surface stations in the US. McIntyre starts dissecting the Jones 1990 paper and his intense focus on individual cases finds a sympathetic ear in Anthony Watts, who launches an even more detailed look at individual cases in the US. Discussions about UHI and data and code turn from a focus on Jones 1990 to a focus on NASA and their GISSTEMP code, which is eventually released. On the FOIA front McIntyre and others are using the tool for a variety of information requests, and most importantly, they use it to get access to the reviewer comments for Chapter 6 of AR4. The pursuit doesn't end there and others pursue FOIA requests for correspondence regarding the writing of IPCC's AR4 Chapter 6. Finally, alerted to the fact that another researcher, Webster, has been given data, McIntyre starts an effort to get global data from Jones. This effort and CRU's attempts to hide behind the act will finally lead to the release of the Climategate files.

The concept of 'an Army of Davids' was popularized by conservative blogger Glenn Reynolds, the Instapundit, in the run-up to the U.S. invasion of Iraq. It refers to the ability of the internet to bring together large numbers of amateur investigators to sift through information available on the internet to shed light on current events, especially scandals. It has been used successfully, often by Pajamas Media, the umbrella organization that now hosts Instapundit, for investigating Democratic misstatements, outright scandals, and in defense of Republican figures and ideas. As many (though certainly not all) skeptics are either independents or Republicans, many of Climate Audit's readers may be veterans of prior expeditions for armies of Davids, and more were certainly aware of the potential for large scale cooperative activities.

With the promise of the release of Jones' 1990 stations, McIntyre pressed on, pointing out problems with Jones 1990, focusing on individual locations and examining what Jones had done in selected cases. Tedious accounting business, but an approach that most readers at Climate Audit found fascinating. One of the persistent conflicts between the Jones camp and the McIntyre camp was the approach to analyzing the UHI problem, making adjustments, and reporting results. Jones 1990 was a 4 page document that described in very general terms the approach that Jones had taken with a few temperature stations. McIntyre and his readers attacked the problem on a case by case basis, focusing on individual sites and examining if Jones' work made sense in these individual cases. It's a task uniquely suited to an army of people looking at individual cases with the same tools.

McIntyre, in a mixture of poking fun at CRU and looking at the detailed numbers, starts the slow process of examining individual cases:

> Low Head is not a manouevre at a Hollywood party or a physical description of the gnomes of Norwich (the location of CRU), but a lighthouse in Tasmania, which John Daly photographed and brought to the attention of Neil Plummer, who had used it in Jones et al 1990. Plummer wrote back to Daly that the inhomogeneity had been taken into consideration in their revised data in Torok and Nichols. Well, was it?
>
> Actually, yes, or at least a qualified yes and the adjustments are pretty interesting. The version from Jones et al 1990 (recently archived) is shown in red. There is a pronounced increase in maximum temperatures, which John Daly had reported to the Australiean BoM as being attributable to changing landscape at the lighthouse. Plummer replied that this had been allowed for in their High Quality network and, as shown in black, they made substantial adjustments to the series. In some years, the adjustment to the maximum temperature is as high as 1.8 degrees C.

Figure 6: Low Head Minimums and Maximums

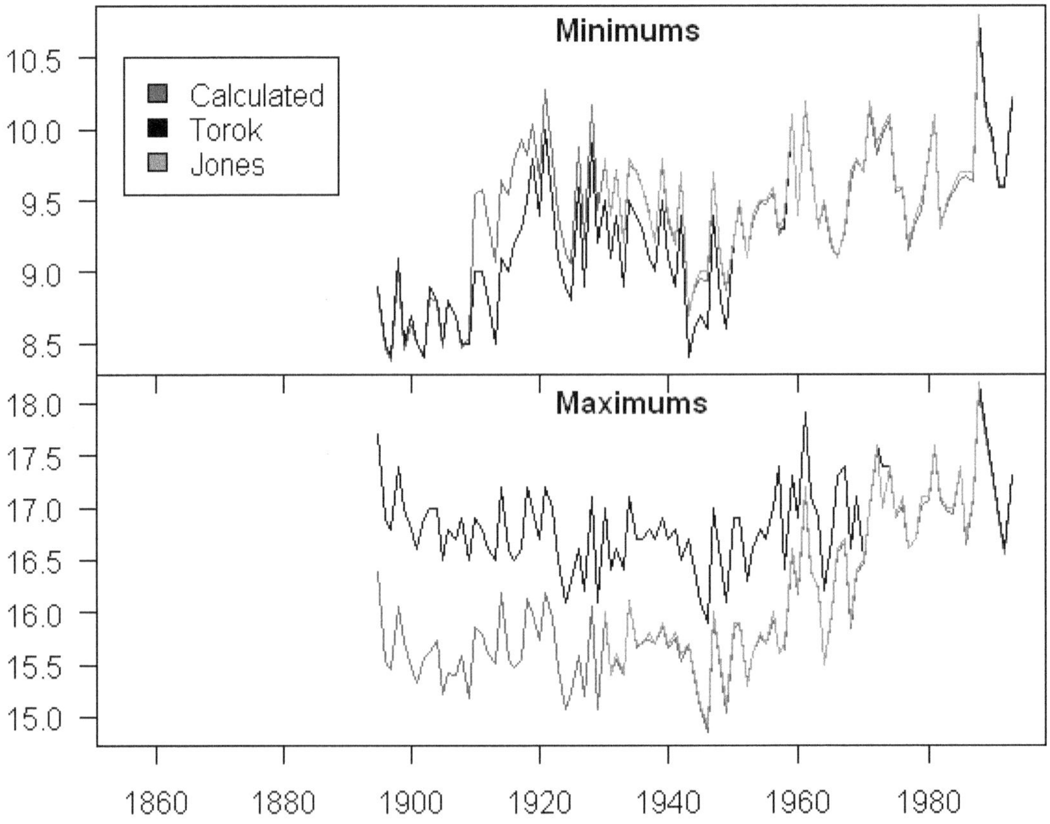

Reproduced by permission, Climate Audit

The discussion of individual stations and regions of the world continues as McIntyre slowly dissects the work done by Jones, trying to understand where the underlying data has come from and what adjustments have been made. Jones' 4 page paper is of little or no help.

By the end of April 2007 the IPCC report AR4 is on line and discussions of climate reconstructions populate the blog. One reconstruction, D'Arrigo and Wilson et al 2007, is of particular interest because it illustrates the way in which the land record and climate reconstructions interact. McIntyre points out to his readers that in one climate reconstruction the authors studying tree rings in Canada have adjusted the temperature record themselves because they thought the official version was inaccurate. Climate reconstructions of the past depend on accurate records from the instrumented period. Here we see authors taking issue with the official record and constructing their own record in order to make better sense of the tree ring data. They don't take the official record as gospel. While climate scientists take no notice of the doubt in the veracity of the record expressed by D'Arrigo and Wilson, they continue to express outrage that McIntyre and his readers would even ask to see the data.

Climate Audit again is on the radar of Phil Jones of CRU and Kevin Trenberth of UCAR. Ben Santer, a former CRU employee, is also alarmed at what is transpiring. Inspired by McIntyre's work, Doug Keenan is reviewing the Jones 1990 case and notices some discrepancies in the work of Jones' co author Wang, a scientist from China. Construing Wang's work as academic fraud, (a charge later dismissed by Wang's university), Keenan starts writing letters to Jones. The case is more interesting for the attitudes and preconceptions it reveals about Jones, Trenberth and others than it is about the questionable work of Wang. Trenberth on April 21 to Jones:

> *I am sure you know that this is not about the science. It is an attack to undermine the science in some way. In that regard I don't think you can ignore it all, as Mike suggests as one option, but the response should try to somehow label these guys and lazy and incompetent and unable to do the huge amount of work it takes to construct such a database. Indeed technology and data handling capabilities have evolved and not everything was saved. So my feeble suggestion is to indeed cast aspersions on their motives and throw in some counter rhetoric. Labeling them as lazy with nothing better to do seems like a good thing to do. How about "I tried to get some data from McIntyre from his 1990 paper, but I was unable because he doesn't have such a paper because he has not done any constructive work!" :*

There is no evidence at all that McIntyre has ever wanted to 'undermine the science.' He's really just asking for the tools to do what they should have done themselves—check their numbers.

Rather than answer Keenan's rather minor issue (which the university will eventually do), Trenberth suggests a personal attack, suggests impugning their motives. Rather than provide the data, Trenberth suggests a false attack on the abilities of those requesting the data. Ironically, the community reading on the internet has far more capability as a whole than Jones at managing large datasets and documents. Many of the engineers, scientists and academics that regularly visit Climate Audit could give lessons on the subject. Trenberth's attack on McIntyre and his readers is reminiscent of the early attacks on bloggers by the mainstream media who tried to label them as "pajamas media." What Trenberth failed to realize is that they were an army of Davids ready to tackle that large database work and in the end provide detailed documentation that NOAA would actually request for its own papers. Mann weighs in with his views and he politicizes the problem and fails to understand that the story has a life of its own, independent of politics and the Main Stream Media:

> *So they are simply hoping to blow this up to something that looks like a legitimate controversy. The last thing you want to do is help them by feeding the fire. Best thing is to ignore them completely. They no longer have their friends in power here in the U.S., and the media has become entirely unsympathetic to the rants of the contrarians at least in the U.S.--the Wall Street Journal editorial page are about the only place they can broadcast their disinformation*

Jones also explains some of the confusion about the actual sources of the data. In CRU FOIA responses, CRU has argued that they should not have to supply the data they used because it is available from GHCN and NCAR. Jones reveals that this is not the case:

> *As for the other request, I don't have the information on the sources of all the sites used in the CRUTEM3 database. We are adding in new datasets regularly (all of NZ from Jim Renwick recently), but we don't keep a source code for each station. Almost all sites have multiple sources and only a few sites have single sources. I know things roughly by country and could reconstruct it, but it would take a while. GHCN and NCAR don't have source codes either. It does all come from the NMSs - well mostly, but some from scientists.*

As has been pointed out earlier, the construction of a global temperature index is largely a book keeping job. Yet Jones seems singularly ill-equipped to keep a handle on the data or accurate records of where data comes from. His breezy assertion that he 'could probably reconstruct' the database record could not be more sharply in contrast to McIntyre's style—or accepted best practice in data handling and management circles. McIntyre's work in the mining industry auditing reports is focused on the issue of provenance. Santer's solution to this "problem" of Climate Audit readers request for data is not a careful exposition of what data sources were used but rather this, from April 25th:

> *I looked at some of the stuff on the Climate Audit web site. I'd really like to talk to a few of these "Auditors" in a dark alley.*

Jones's best thought at clearing up this confusion in data sources and data provenance is to deliberately confuse the matter more. He writes:

> *Ben,*
>
> *Thanks for the thoughts. I'm in Geneva at the moment, so have a bit of time to think. Possibly I'll get the raw data from GHCN and do some work to replace our adjusted data with these, then make the Raw (i.e. as transmitted by the NMSs). This will annoy them more, so may inflame the situation.*

Reviewing the ideas the scientists have for dealing with data requests we have the following: Lie about the dedication and work habits of the requester, ignore requests, beat requesters up in a dark alley, and deliberately confuse the provenance of the data. Later they will add obstruction of the FOIA process to their bags of tricks, demonstrating that they should never in the future be trusted with data that is fundamental to our understanding of the climate. Releasing the data and letting the "auditors" discredit themselves never occurs to them.

A quick digression—your co-authors, Steven Mosher and Thomas Fuller, are quite different politically. Fuller describes himself as a progressive liberal who has never voted for anybody who was not a Democrat. But it was reading the above sequence of emails that convinced Fuller that this book was a needed corrective to the way things have been done regarding climate science in general and data hygiene in particular. If these premiere-level climate scientists are willing to play games at this level with bloggers, what games are they willing to play to win larger and more serious contests?

At the start of May, McIntyre links to a blogger named Anthony Watts, a former TV meteorologist who was convinced that temperature monitoring stations in the United States were in dire shape and could not be trusted to create a temperature record, especially one that the world would use as a reference point for dealing with climate change. During the summer, Watts would launch a nationwide volunteer effort to document the weather collection stations used by NOAA, NASA, CRU and Jones. The effort that Trenberth thought too large for any one individual would be handled under Watts' generalship by a true army of Davids across the nation, using the tools of the internet. The goal very simply was to document the status of the temperature collection stations. Many hands made light work of the job scientists thought too large to attempt.

Tom Karl of NOAA takes notice of Watts but is not sure how it will turn out.

> *From: "Thomas.R.Karl" <Thomas.R.Karl@noaa.gov>*
>
> *To: Phil Jones <p.jones@uea.ac.uk>*
>
> *Subject: Re: FW: retraction request*
>
> *Date: Tue, 19 Jun 2007 08:21:57 -0400*
>
> *Thanks Phil, We R now responding to a former TV weather forecaster who has got press, He has a web site of 40 of the USHCN stations showing less than ideal exposure. He claims he can show urban biases and exposure biases. We are writing a response for our Public Affairs. Not sure how it will play out. Regards, Tom*

That effort, ridiculed at first by bloggers in the warmist faction, would in the end garner Watts a visit to NCDC to discuss his work. Moreover, NOAA would engage in an effort to bring the climate network up to better quality standards. As of July 2009 the volunteer effort, hosted at wwwsurfacestations.org. had surveyed 1,003 of the 1,221 stations used by NOAA and corrected mistakes in the official metadata.:

We here take a look at some of what they found—it's a bit shocking, really. Only 11% of temperature monitoring stations in this country meet guidelines set up to insure they accurately collect and report temperature.

Figure 7: Temperature Collection Stations Audited by Watt's Up With That Volunteers

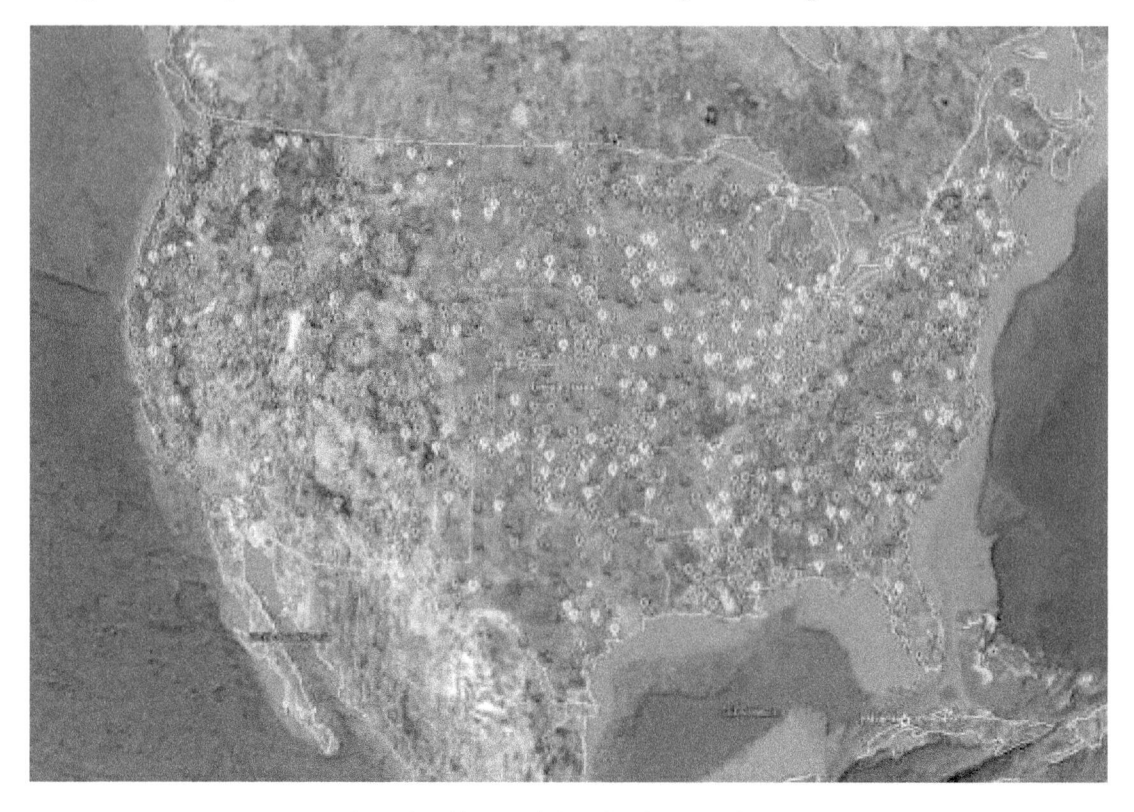

Reproduced by permission, Watt's Up With That

Watts and his volunteers documented the problems with each site—problems that may cause issues with the accuracy of temperature measurements. A typical example of a site with siting problems is shown below, one of Watts' first finds, in Marysville California.

Figure 8: Temperature Collection Station

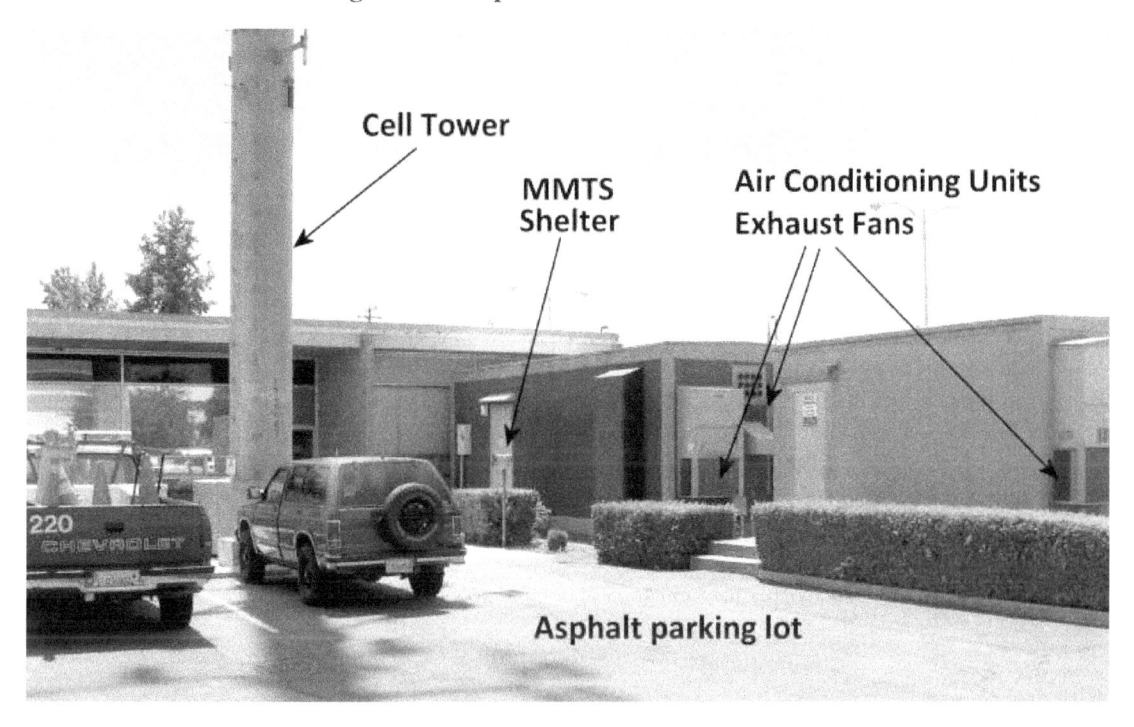

Reproduced by permission, Watt's Up With That

Watts's work goes to the core of one of the other key papers on UHI relied on by the IPCC: Peterson's 2003 paper on UHI. Peterson's paper, like Jones', investigated whether or not the historical record was infected by UHI. To do this, Peterson, like Jones, selected stations he believed were rural and compared them to stations he believed were urban. The basis of categorization was a satellite photo of the nighttime lighting at each site. While a photo of "nightlights" can identify a site that is located in a well lit and presumably urban area, the absence of lights at night is no guarantee that the site is rural and pristine. That would require close up detailed examination of each site. Field work as opposed to office work. Watts and his volunteers would do what Peterson would not or could not.

In order to understand the importance of this work one must understand the basic argument that Peterson, and later Parker and Jones, would make. From Peterson 2003:

1. Introduction

a. Impetus for this analysis

As just about every introductory course on weather and climate explains, urban areas are generally warmer than nearby rural areas. Often referred to as the urban heat island (UHI) effect, urbanization has long been regarded as a serious contamination of the climate signal (e.g., Landsberg 1956). Those of us working with century-scale instrumental climate data strive to remove all sources of artificial biases from the data. So the UHI contamination is one aspect dataset creators seek to address. For example, the Global Historical Climatology Network (GHCN; Peterson and Vose 1997) consists of over 7,500 temperature stations around the world that were identified as rural, urban, or an in-between class of small town

> using information on operational navigation charts and a variety of different atlases. A rural station was any station not associated with a town of over 10,000 population.

Jones and CRU claim that urban stations have warmed .05C per century more than rural. When Peterson 2003 found **no difference** between the urban and the rural he declared the finding a mystery.

> **The linear trend from 1880 to 1998 was 0.65C century for the full dataset and the slightly higher 0.70C century for the rural-only sub-set. The resulting conclusion was that the well-known global temperature time series from in situ stations was not significantly impacted by urban warming. The research presented here attempts to unravel the mystery of how a global temperature time series created partly from urban in situ stations could show no contamination from urban warming. This is important to improving our understanding of the UHI contamination of in situ temperature observations and therefore the fidelity of the climate change measurements. This point is highlighted by the fact that some "greenhouse skeptics" continue to argue that a significant portion of the observed warming is only an urban effect (Hansen et al. 2000).**

In reviewing the literature on UHI, Peterson cites Wang, Jones' co author: Wang had found a trend in urban stations that was greater than rural stations but more importantly Wang had found that rural stations were not true rural stations.

> **One approach, the time rate of change method, looked at differences in temperature time series between urban and rural. The results of this method indicated that the warming rate using 42 pairs of urban–rural stations in China, Wang et al. (1990) found an average urban heat island of 0.23C "despite the fact that the rural stations were not true rural stations." "Multiple regression techniques" were used "to minimize the effects of differences in altitude, latitude and longitude." No details or additional information were provided on this or any other aspect of homogeneity.**

One problem Peterson noted with the metadata approach is that there was no way to ascertain if it was actually correct. Just because the metadata says a station is rural is no guarantee that it is rural, as Wang had noted. So Peterson relied on an approach used by James Hansen. Look at a satellite photo of the site at night and determine if it was rural by looking at the intensity of lights at night. In Peterson's mind this type of "picture" is good enough to establish if the rural site is really "rural."

There were other potential biases that Peterson sought to remove. Most importantly Peterson considered the question of micro-siting bias. In other words, how do you get a closer look at the site, closer even than a picture of its lights from an orbiting satellite?

> **Microscale siting characteristics can produce biases in the temperature measurements. Assessing these characteristics can be both extremely difficult, given the level of available metadata, and is part of the essential rural/urban question this analysis seeks to address. However, one siting characteristic that is not part of the rural/ urban question and that can impart a large bias is non- standard siting, particularly rooftop observations. Roof top observations tend to be warmer at night due to being higher in the stably stratified nocturnal boundary layer and warmer**

during the day due to less thermal mass below them being warmed by the sun and less available water to be converted into latent heat. This problem has been known for decades. Indeed, Landsberg (1942) states that the differences between rooftop and ground-based observations "indicates clearly that conclusions on climate derived from records of roof stations may by no means be representative of those at the ground level. Stations located at roof level and on tall buildings have been used in the past. Effects of site changes before the move was made resulted in some valuable site-specific information. The results of the Davey et al. (2002) analysis of rooftop stations indicated that rooftop sites are usually warmer than nearby ground-based observations.

Peterson notes the difficulty of this problem lies in the quality of metadata, the data in the file that describes the station. If the metadata indicated the station was on a rooftop it was removed from Peterson's analysis. Two such stations were found. It never occurred to Peterson to inspect each and every one of the sites. He inspected the metadata and trusted it. The "bottoms up" approach, the approach of actually visiting each site, the approach taken by Watts, will not work for the scientist working from an office. It would require field work, time and effort—and money. Later Watts' volunteers, constituting a team that Trenberth wanted to label as lazy, would find stations in places more odd than rooftops: They would produce data not captured in metadata files:

Figure 9: Temperature Collection Station

Reproduced by permission, Watt's Up With That

The conclusion Peterson came to was that the urban stations had warmed no more than the rural stations, once the factors of instrument changes and microsite issues (two rooftop stations) were removed. But note that the only microsite issue that concerned Peterson was rooftop placement. He could only correct for this error because the metadata he had was limited.

The results, the lack of urban warming, were a mystery to Peterson because there **should** be a difference between urban and rural, as he states in his opening paragraph. Peterson attempts to solve the mystery in his concluding paragraphs and he turns to the literature:

> **Recent research by Spronken-Smith and Oke (1998) also concluded that there was a marked park cool island effect within the UHI. They report that under ideal conditions the park cool island can be greater than 5 degrees C, though in mid-latitude cities they are typically 1–2C. In the cities studied, the nocturnal cooling in parks is often similar to that of rural areas. They reported that the thermal influence of parks on air temperatures appears to be restricted to a distance of about one park width.**
>
> **Therefore, if a station is located within a park, it would be expected to report cooler temperatures than the industrial sections experience. But do the urban meteorological observing stations tend to be located in parks or gardens? The official National Weather Service guidelines for non-airport stations state that an observing shelter should be "no closer than four times the height of any obstruction (tree, fence, building, etc.)" and "it should be at least 100 feet from any paved or concrete surface" (Observing Systems Branch 1989). If a station meets these guidelines or even if any attempt to come close to these guidelines was made, it is clear that a station would be far more likely to be located in a park cool island than an industrial hot spot.**

We can summarize Peterson as follows. Peterson **believes** that the siting of stations in urban environments follows siting guidelines. That is, he believes they are 100 feet from any paved or concrete surface. If a site is properly sited, then it is likely that it is located in what Oke called a "cool park." We also see something of Peterson's approach in this matter. The official literature says that stations **should be** sited properly, therefore it is likely that **they are** properly sited. Peterson trusts official records. And if they are properly sited, the literature also says they are likely to be located in cool parks. He trusts the literature. Despite his initial surprise, Peterson then concludes that urban sites are no different than rural. Urban sites must be in cool parks.

> **It is postulated that the reason for this is due to micro- and local-scale impacts dominating over the mesoscale urban heat island. Industrial sections of towns may well be significantly warmer than rural sites, but urban meteorological observations are more likely to be made within park cool islands than industrial regions.**

It should be noted that Peterson's contention about cool parks is stated as a postulate. Peterson also notes that the rural sites could be contaminated. And so the mystery is "solved" if one accepts the postulates that urban sites are located in "cool parks" and rural sites may not always be that pristine. So, unlike a famous English detective who solved a case of the 'dog that didn't bark,' Peterson assumes all is well, even though he cannot find the urban heat effect that he knows should be there.

Armed with NOAA's own guideline for siting stations, Watts' volunteers fan out across the country to document whether Peterson's postulates held any water. Rather than relying on literature to determine if sites were actually sited properly, the volunteers actually visited the sites to photo document compliance with the code. Were sites actually located in cool parks, as Peterson postulated, or were they located near to and sometimes over asphalt? The results were embarrassing to NOAA and for a brief time they blocked Watts' access to the location data his volunteers needed to visit the sites. That shocking action is hard to fathom. Watts was building a nationwide volunteer program to visit every site and photo document it. According to NOAA's own requirements, photo documentation was required. But NOAA had not done this. Faced with a nationwide volunteer effort to do the basic field work which scientists had not done, the scientists' reaction was to shut down access to the data Watts needed to actually visit the sites. That obstruction was eventually removed and the volunteers posted a hall of shame for NOAA. It numbered in the hundreds.

Figure 10: Temperature Collection Station

Reproduced by permission, Watt's Up With That

Figure 11: Temperature Collection Station

Reproduced by permission, Watt's Up With That

In the end, Watts would apply NOAA's own site rating guide to the sites his volunteers had photo documented, assigning each a rating according to a protocol endorsed by NOAA. Urban sites did not occur in cool parks and rural sites were more contaminated than Peterson imagined, and the problems with sites were not limited to the problem of rooftop siting: Roughly 11% of all stations met guidelines.

Santer, writing in defense of Jones and his work, appears to miss the importance of the documentation the volunteers are performing.

> *They conveniently ignore all the pioneering work that you've done on identification of inhomogeneities in surface temperature records. The response should mention that you've spent much of your scientific career trying to quantify the effects of such inhomogeneities, changing spatial coverage, etc. on observed estimates of global-scale surface temperature change. The bottom line here is that observational data are frequently "messy". They are not the neat, tidy beasts .We would like observing systems to be more accurate, more stable, and better-suited for monitoring decadal-scale changes in climate. You and Kevin and many other are actively working towards that goal. The key message here is that, despite uncertainties in the surface temperature record - uncertainties which you and others in the field are well aware of, and have worked hard to quantify - it is now unequivocal that surface temperatures have warmed markedly*

108

Perhaps because he is overly focused on what the proper 'message' should be, what Santer misses is that these people are not ignoring the pioneering work of Jones and scientists like Peterson. Rather they are examining the very postulates that Peterson, and later Parker and later Jones would make about siting in "cool parks."

Figure 11: Temperature Station Quality Percentages

USHCN - Station Site Quality by Rating

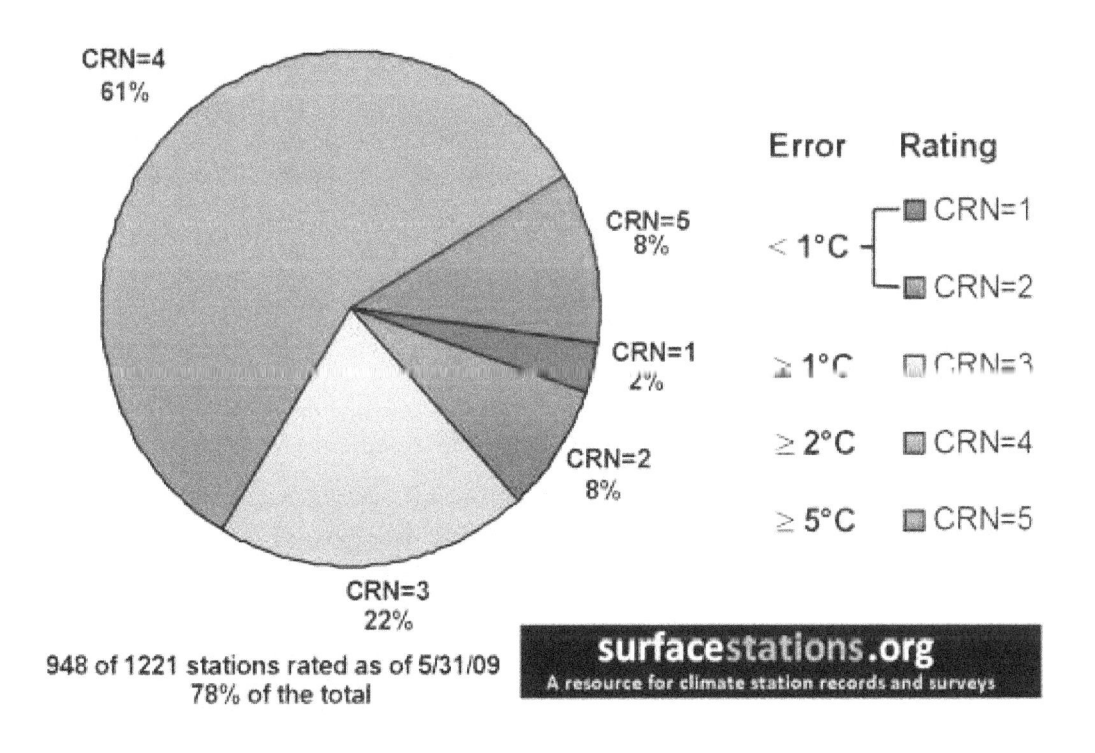

- CRN=4 61%
- CRN=5 8%
- CRN=1 2%
- CRN=2 8%
- CRN=3 22%

Error	Rating
< 1°C	CRN=1
	CRN=2
≥ 1°C	CRN=3
≥ 2°C	CRN=4
≥ 5°C	CRN=5

948 of 1221 stations rated as of 5/31/09
78% of the total

surfacestations.org
A resource for climate station records and surveys

Reproduced by permission, Watt's Up With That

At the time of this writing, Watts has a paper in presentation for publication. When interviewed he revealed, "Two papers are planned with Dr. Roger Pielke of the University of Colorado and other authors. We have already done the analysis, and have discovered a distinct signal showing the difference between poorly sited stations and well sited ones. The difference is most prominent in the nighttime low temperature data, as would be expected, since so many climate monitoring stations have been found next to heat sinks like asphalt, concrete, and buildings."

Auditing the IPCC process

On a different front McIntyre is pursuing issues with the IPCC. As a published author in the field of climate reconstructions, we saw how McIntyre had been asked to review Chapter 6 of AR4. Briffa, who has refused in the past to supply data to McIntyre, was the lead author of the contentious section. McIntyre who is eager to write about how Briffa has brushed off several of his concerns about "hiding the decline" and other matters, inquires as to when the IPCC will make reviewers' comments public as they have promised to do. The IPCC responds by stating that the reviewer comments are available in Harvard's Public Policy Archives and gives McIntyre a notice of the visiting hours. McIntyre can visit his own comments and the comments of others in the library. It is as if the IPCC had not discovered the internet. A policy that might have been— barely—adequate when the IPCC was formed had not moved with the times.

Since the IPCC was supposed to be dedicated to an open and transparent process, McIntyre responds by issuing FOIA's to NOAA and Susan Solomon. Eventually the organization is forced to post the comments on the web, but not before they try another half-compliant method to answer McIntyre's request for the reviewer comments. They offer to send him copies but not allow him to redistribute them. Not exactly a model of openness and transparency.

At Climate Audit, McIntyre writes,

IPCC has just written me saying that they will send me review comments on Chapter 6 subject to the following restriction:

As this additional form of distribution is being provided in conjunction with the review process, the compiled comments are not for re-distribution to others.

Given that the review comments are supposedly in an "open archive", I don't understand the basis of this restriction. Also I'm not clear whether this prohibits me from quoting even individual review comments. It's all very strange and very inconsistent with the "open and transparent process" that IPCC is supposed to follow. Much as bureaucratic obfuscation amuses, even I'm getting tired of WG1 TSU, so I've tried this from a different angle.

Many of the key players in WG1, including the Chairman, Susan Solomon, and the TSU director, Martin Manning, are NOAA employees ...

In the wake of Climategate many of The Team's defenders have complained about the FOIA requests made by McIntyre and his readers, saying they were burdensome and distracting. Those complaints do not appear to us to be well-grounded. The record shows that the IPCC was dedicated to an open and transparent review process. And in almost all the resolved FOIA issues, once the responsible authority decided to grant the information requests, the information was easily located and made available, sometimes within days.

McIntyre made a simple request that could have been easily met. Instead, the IPCC put up obstacles. So McIntyre proceeded to issue an FOIA request to NOAA at the end of May 2007. Jones is apprised of McIntyre's move and in an email to Tom Karl, Jones comments on this as well as his efforts to get UEA to ignore FOIA.

1. Think I've managed to persuade UEA to ignore all further FOIA requests if the people have anything to do with Climate Audit.

2. Had an email from David Jones of BMRC, Melbourne. He said they are ignoring anybody who has dealings with CA, as there are threads on it about Australian sites.

3. CA is in dispute with IPCC (Susan Solomon and Martin Manning) about the availability of the responses to reviewer's at the various stages of the AR4 drafts. They are most interested here re Ch 6 on paleo.

Jones' efforts within CRU to have requests by readers of Climate Audit viewed differently appears to be at odds with accepted policies and procedures. The goal of FOIA is to release information to the public regardless of their views on particular matters. A political party in power, for example, cannot deny an FOIA based on the political views of the people requesting data. It is unclear whether the internal CRU investigation being conducted at the time of this writing will investigate this. At McIntyre's urging four of his readers join him in his FOIA requests and before the deadline to respond expires NOAA works to get the reviewer comments posted online.

GISSTEMP: The Sacrificial Lamb

Anthony Watt's volunteers continue to send in photos of NOAA's temperature stations. The picture below, in fact leads to a short battle between Climate Audit and Real Climate over NASA's version of the global temperature index.

Figure 12: Temperature Collection Station

Reproduced by permission, Watt's Up With That

In climate science there are two notable records of the global temperature. HADCRUT, maintained by Jones, and NASA's GISSTEMP. Since Watt's focus is on the US temperature stations, he focuses on GISSTEMP. Stymied by CRU and inspired by Watts, McIntyre turns his attention to GISSTEMP. In short order, McIntyre finds a glaring error in NASA's processing, and a campaign to "free the code" is launched. NASA relents and publishes their code in September of 2007: This event is noted by Jones who writes on Sept 11th,

> *PS to Gavin - been following (sporadically) the CA stuff about the GISS data and release of the code etc by Jim [Hansen]. May take some of the pressure of you soon, by releasing a list of the stations we use - just a list, no code and no data. Have agreed to under the FOIA here in the UK.*

On Oct 1 2007, as promised, CRU reveals their list of stations. No data. No code to tell people how the temperature index is created. Jones just releases the names of stations. As the year closes out McIntyre's readers collect money and send him to a conference in San Francisco where he meets with Watts and gives a presentation. As the year closes out Climate Audit wins an award, Science Blog of the Year, and McIntyre introduces readers to a new blog called The Blackboard, run by Lucia Liljegren. Two years later, the Climategate scandal will break in a post on her site.

As 2008 started, Climate Audit still had some issues left on the table, such as CRU data, which would not go away with the new year. Finally, McIntyre's pressure on Solomon yielded fruit and the reviewer comments to AR4 were put online and readers pored over comments, noting which were accepted, which were rejected and which were flatly ignored. The focus on NASA and nightlights used to judge the rural quality of stations drew attention and McIntyre did a series of technical posts on topics like Toeplitz Matrices and tree ring networks and spatial autocorrelation. Of your authors, Mosher claims to understand them. Fuller does not.

Interest in the comments of various reviewers of AR4 increased and one CA reader, David Holland, made it a point to launch into his own FOIA inquiry into the comments made by MET head scientist John Mitchell. Mitchell's initial reply was that he did not keep any correspondence or working papers. Holland persisted in his inquiries over the following months requesting the correspondence of Briffa and Caspar Ammann. The correspondence of the last two was especially interesting to McIntyre because he wanted to solve a puzzle about the acceptance of the Ammann paper, the "Jesus Paper," (as it was dubbed by Bishop Hill, an English blogger) after the AR4 deadlines had passed.

By carefully comparing the Lead Authors' comments (Briffa) with Ammann's paper, which was published after Briffa made the comments, it appeared clear to McIntyre that Briffa and Ammann had discussed the unpublished paper off the record. As we saw from the mail Briffa sent to Wahl, Ammann's co author, marked confidential, Briffa and the author's of the 'Jesus Paper' were discussing more than the paper. Briffa was passing comments to Wahl, asking for his help in maintaining his "objectivity"

Inside CRU and NOAA, McIntyre and Holland had raised hackles and tempers grew short and Jones focuses on methods to avoid compliance with FIOA requests:

> *Jones : 2. You can delete this attachment if you want. Keep this quiet also, but this is the person[Holland] who is putting in FOI requests for all emails Keith and Tim have written and received re Ch 6 of AR4. We think we've found a way around this.*

And Tim Osborne writes Ammann, suggesting a possible way to avoid the requests. Essentially, we see two scientists working on Chapter 6 of AR4. That document is supposed to be constructed in an open and transparent process. Holland requested the correspondence of Briffa and Osborne, which would have included the mail Briffa sent to Wahl and the mail Wahl sent to Briffa. Emails outside the process which point to the very special status and treatment that paper got. Rather than simply comply, Jones looks for a way around the request.

> *Dear Casper[Ammann],*
>
> *I hope everything's fine with you. Our university has received a request, under the UK Freedom of Information law, from someone called David Holland for emails or other documents that you may have sent to us that discuss any matters related to the IPCC assessment process. We are not sure what our university's response will be, nor have we even checked whether you sent us emails that relate to the IPCC assessment or that we retained any that you may have sent. However, it would be useful to know your opinion on this matter. In particular, we would like to know whether you consider any emails that you sent to us as confidential. Sorry to bother you with this,*

On advice from Palmer that they can avoid the requests if the material was considered confidential, Jones feeds Ammann the answer he needs. Note that neither Jones nor Palmer informs Ammann about the guidelines IPCC documents and processes must follow. And Jones adds another mechanism to avoid releasing the material and he seems to be directing the answers his team should give. The dodge works as follows. In the IPCC process the authors are supposed to consider work that is published in the peer reviewed journals. The reviewers are likewise supposed to be given access to this literature.

What McIntyre and Holland were concerned about was communication outside the IPCC process. Communication that would subvert the open and transparent process of literature review. As Jones points out, the method used to subvert the open and transparent process was merely to send papers directly to authors. That way the final document could be influenced but not through official channels, such as reviewer comments. So Jones tells Palmer, The FOIA officer, that Briffa should just say he didn't get any comments that didn't come through the normal process. This is clearly not true.

The director of the Climate Research Unit, Phil Jones, is advising the lead editor of the IPCC's AR4 Chapter 6 with the complicit knowledge of the UK Freedom of Information Act's officer responsible for compliance with its provisions that they should lie about provisions of the report that would inform climate change policy across the world.

> *From: Phil Jones <p.jones@uea.ac.uk>*
>
> *To: t.osborn@uea.ac.uk,"Palmer Dave Mr \(LIB\)" <David.Palmer@uea.ac.uk>*
>
> *Subject: Re: FW: Your Ref: FOI_08-23 - IPCC, 2007 WGI Chapter 6 Assessment Process [FOI_08-23]*
>
> *Date: Wed, 28 May 2008 17:13:35 +0100*
>
> *Cc: "Briffa Keith Prof\" <k.briffa@uea.ac.uk>, "Mcgarvie Michael Mr\" <m.mcgarvie@uea.ac.uk>*
>
>> Dave,
>>
>> *Although requests (1) and (2) are for the IPCC, so irrelevant to UEA, Keith (or you Dave) could say that for (1) Keith didn't get any additional comments in the drafts other than those supplied by IPCC. On (2) Keith should say that he didn't get any papers through the IPCC process either. I was doing a different chapter from Keith and I didn't get any. What we did get were papers sent to us directly - so not through IPCC, asking us to refer to them in the IPCC chapters. If only Holland knew how the process really worked!! Every faculty member in ENV and all the post docs and most PhDs do, but seemingly not Holland. So the answers to both (1) and (2) should be directed to IPCC, but Keith should say that he didn't get anything extra that wasn't in the IPCC comments.*
>>
>> *As for (3)[the communication is "confidential"] Tim has asked Caspar, but Caspar is one of the worse responders to emails known. I doubt either he emailed Keith or Keith emailed him related to IPCC. I think this will be quite easy to respond to once Keith is back. From looking at these questions and the Climate Audit web site, this all relates to two papers in the journal Climatic Change. I know how Keith and Tim got access to these papers and it was nothing to do with IPCC.*

Cheers   Phil

As that mail reveals, Keith Briffa, lead author of Chapter 6 of AR4, did have access to the papers that McIntyre was wondering about, but not through any "IPCC" process. The papers were sent directly to Briffa and Osborne. In any case it appears that McIntyre's concern about Briffa making lead author decisions based on Ammann's unpublished works was correct. But the whole defense that this correspondence was confidential was laid out **before** Jones even contacted people. The mails also show FOIA officer David Palmer preemptively laying out the defense on May 27th

> *Could you*
>
> > *provide input as to his additional questions 1, and 2, and check with Mr. Ammann in question 3 as to whether he believes his correspondence with us to be confidential? Although I fear/anticipate the response, I believe that I should inform the requester that his request will be over the appropriate limit and ask him to limit it - the ICO Guidance states: 12. If an authority estimates that complying with a request will exceed the cost limit, can advice and assistance be offered with a view to the applicant refocusing the request? In such cases the authority is not obliged to comply with the request and will issue a refusal notice. Included within the notice (which must state the reason for refusing the request, provide details of complaints procedure, and contain particulars of section 50 rights) could be advice and assistance relating to the refocusing of the request, together with an indication of the information that would be available within the cost limit (as required by the Access Code). This should not preclude other 'verbal' contact with the applicant, whereby the authority can ascertain the requirements of the applicant, and the normal customer service standards that the authority usually adopts. And.. our own Code of Practice states (Annex C, point 5) 5. Where the UEA is not obliged to supply the information requested because the cost of doing so would exceed the "appropriate limit" (i.e. cost threshold), and where the UEA is not prepared to meet the additional costs itself, it should nevertheless provide an indication of what information could be provided within the cost ceiling. This is based on the Lord Chancellors Code of Practice which contains a virtually identical provision. In effect, we have to help the requester phrase the request in such a way as to bring it within the appropriate limit - if the requester disregards that advice, then we don't provide the information and allow them to proceed as they wish. I just wish to ensure that we do as much as possible 'by the book' in this instance as I am certain that this will end up in an appeal, with the statutory potential to end up with the ICO.*

Palmer plans to argue that the request will cost too much and the correspondence is "confidential" prior to investigation. Palmer signals that he wants to run everything "by the book" as an appeal is likely. Palmer's desire to respond to this request by the book, of course, raises the question of why he has to make this distinction clear to Jones. In short, prior to this Jones has been instrumental in directing the responses of the FOIA office. In this case, Palmer feels he must run things by the book. It is in this context that Phil Jones wrote to Michael Mann asking him to delete any mails he has with Briffa with regard to Chapter 6 of AR4 on May 29 2008

To recap—McIntyre publishes a critical paper (MM05) in February of 2005. From that time until AR4 Chapter 6 is finished, the Team is struggling to get Ammann's paper into print, because Ammann's paper will give Briffa a source in peer reviewed literature he can use to blunt

McIntyre's criticism. They want to remove the doubt that McIntyre raises. McIntyre throughout the whole process is watching Ammann's paper struggling coming to press and he documents this struggle in real time on Climate Audit. The reviewers' comments on Chapter 6 are freed and it's apparent that Briffa has access to Ammann's paper. McIntyre and his readers go after the correspondence of the authors. Jones and Palmer believe they can refuse the request by citing confidentiality, but Jones needs to worry about an appeal. He asks Mann to direct Wahl, Ammann's co author, to delete mails.

The defense of confidentiality is created presumptively prior to investigation. And while they are in the process of asking Ammann if his mails are confidential at Palmer's direction, Jones tells people to delete their mails. Which begs the question: If he believed in the defense Palmer set out, what was the point of deleting mails? Was it Palmer's insistence that this be done "by the book" or the fear of appeal or both?

> *Mike, Can you delete any emails you may have had with Keith re AR4? Keith will do likewise. He's not in at the moment - minor family crisis. Can you also email Gene[Wahl] and get him to do the same? I don't have his new email address. We will be getting Caspar to do likewise…*
>
> *Phil*

Their abuse of the FOIA process has become so commonplace that this incredible request is preserved in the record.

On May 30th, Osborne reiterates his request to Ammann, prompting him for the answer Palmer has predetermined. Osborne gives Ammann the answer they want to hear back from him.

> *From: Tim Osborn <t.osborn@uea.ac.uk>*
>
> *To: Caspar Ammannn <Ammannn@ucar.edu>*
>
> *Subject: Re: request for your emails*
>
> *Date: Fri May 30 12:58:34 2008*
>
> *Cc: "keith Briffa" <k.briffa@uea.ac.uk>, p.jones@uea.ac.uk*
>
> *Hi again Caspar, I don't think it is necessary for you to dig through any emails you may have sent us to determine your answer. Our question is a more general one, which is whether you generally consider emails that you sent us to have been sent in confidence. If you do, then we will use this as a reason to decline the request.   Cheers   Tim*

And Ammann responds a bit confusedly on May 30th.

Hi Tim, in response to your inquiry about my take on the confidentiality of my email communications with you, Keith or Phil, I have to say that the intent of these emails is to reply or communicate with the individuals on the distribution list, and they are not intended for general 'publication'. If I would consider my texts to potentially get wider dissemination then I would probably have written them in a different style. Having said that, as far as I can remember (and I haven't checked in the records, if they even still exist) I have never written an explicit statement on these messages that would label them strictly confidential. Not sure if this is of any help, but it seems to me that it reflects our standard way of interaction in the scientific community. Caspar

Ammann has no more clarification, yet on June 3, Palmer issues his denial, citing the exceptions he had preordained the group would follow. Palmer predetermined that the excuse would be 'confidentiality." He asked for Ammann's view. Ammann responded confusingly and Palmer proceeds with his excuse regardless: he argues that they have always considered that the mails are confidential, that Ammann considers them to be confidential and it will cost too much to comply. Whoever collected the Climategate mails apparently differed with Mr. Palmer's cost of compliance "assessment."

Dear Mr. Holland,

FREEDOM OF INFORMATION ACT 2000 – INFORMATION REQUEST (Our Ref: FOI_08-23)

Your request for information received on 5 May 2008 has now been considered and it is, unfortunately, not possible to meet your request.

In accordance with s.17 of the Freedom of Information Act 2000 this letter acts as a Refusal Notice, and I am not obliged to supply this information and the reasons for exemption are as stated below

Exemption and Reason

s.12, Cost of compliance exceeds appropriate limit: The cost of finding & assembling the information will exceed the appropriate limit

s.41, Information provided in confidence: The release of this information would constitute an actionable breach of confidence

Given the amount of material covered by your request, the cost of compliance in locating, retrieving and in the reading, editing or redaction of the relevant documents would clearly exceed the appropriate limit.

Additionally, we hold that the s.41 exemption applies to all requested correspondence received by the University. We have consistently treated this information as confidential and have been assured by the persons and organisations giving this information to us that they believe it to be confidential and would expect to be treated as such.

> **The public interest in withholding this information outweighs that of releasing it due to the need to protect the openness and confidentiality of academic intercourse prior to publication which, in turn, assures that such cooperation & openness can continue and inform scientific research and debate.**
>
> **I apologise that your request will not be met but if you have any further information needs in the future then please contact me.**
>
> *David Palmer*

This refusal, of course raises the following question: If CRU have always considered these mails to be confidential, then why would Jones advise people to delete mails? True confidential status would be a valid reason for denial. The request to delete mails makes sense only if Jones and Palmer believed an appeal was likely, and only if they believed an appeal was likely to be granted. That is, CRU had never considered such mails to be confidential. And finally, the request to delete mails only makes sense if the mails held information likely to be damaging.

After the denial is sent to Holland in early June, Briffa on June 23rd puts his position "on the record" with Ammann. Briffa, it seems, is trying to create a paper trail of his position in case there is an appeal. By June 23 Holland's request has already been denied. Palmer requested that Ammann be asked his position at the end of May, a couple days after Ammann is asked and replies in a confused way, Palmer denies the request on June 3. Twenty days later Briffa writes,

Date: Mon Jun 23 09:47:54 2008

> *Caspar I have been of the opinion right from the start of these FOI requests, that our private, inter-collegial discussion is just that - PRIVATE. Your communication with individual colleagues was on the same basis as that for any other person and it discredits the IPCC process not one iota not to reveal the details. On the contrary, submitting to these "demands" undermines the wider scientific expectation of personal confidentiality. It is for this reason, and not because we have or have not got anything to hide, that I believe none of us should submit to these "requests". Best wishes Keith*

Briffa is obviously taking issue with Ammann's position on the openness and transparency requirements of the IPCC. Briffa argues that personal confidentiality trumps the public's right to know. The scientist is more important than the public. Osborne, alert to the conflict between Ammann and Briffa, argues for consistency of action. They all have to behave as if these documents are confidential. And this is after Palmer has denied the request on the grounds that the correspondence is confidential. Apparently Palmer asked for Ammann's opinion and did not take notice of it. Osborne appears to have received advice about this matter and presses Ammann.

> *At 09:01 23/06/2008, Tim Osborn wrote:*
>
> > *Hi Caspar, I've just had a quick look at CA. They seem to think that somehow it is an advantage to send material outside the formal review process. But *anybody* could have*

emailed us directly. It is in fact a disadvantage! If it is outside the formal process then we could simply ignore it, whereas formal comments had to be formally considered. Strange that they don't realise this and instead argue for some secret conspiracy that they are excluded from! I'm not even sure if you sent me or Keith anything, despite McIntyre's conviction! But I'd ignore this guy's request anyway. If we aren't consistent in keeping our discussions out of the public domain, then it might be argued that none of them can be kept private. Apparently, consistency of our actions is important. Best wishes Tim

Holland appeals, arguing that the IPCC process was supposed to be open, but he is denied. Strangely, MET which works in conjunction with CRU, changed their reason for denying Holland's request for Mitchell's correspondence. At first Mitchell claimed to have no working papers and claimed they were lost. In the end MET adopted CRU's dodge and argued that Mitchell's contributions to the IPCC were personal.

As July came in so did a notice from NOAA. A FOIA that McIntyre had sent on April of 2007 was lost, this being the request for the Jones 1990 stations and data. So finally after CRU decided to release that data, NOAA found its paperwork and sent McIntyre the data:

> **As previously indicated in my e-mail last Thursday, NCDC did indeed put together a response to your official FOIA request, but due to a miscommunication between our office and our headquarters, the response was not submitted to you. I deeply apologize for this oversight, and we have taken measures to ensure this does not happen in the future. Attached for your reference is the actual response to your inquiry that was provided on April 4, 2007. Please let me know if I or my office may assist you further. Thank you.**
>
> **Sincerely,**
>
> **Tom Karl**

In an August mail to Gavin Schmidt, Jones explains his activities with the FOIA officers and their pre-ordained strategies to not respond.

> *but I don't want to give them something clearly tangible. Keith/Tim still getting FOI requests as well as MOHC and Reading. All our FOI officers have been in discussions and are now using the same exceptions not to respond.... The FOI line we're all using is this. IPCC is exempt from any countries FOI - the skeptics have been told this. Even though we (MOHC, CRU/UEA) possibly hold relevant info the IPCC is not part our remit (mission statement, aims etc) therefore we don't have an obligation to pass it on.*

The End of the World

The end of September brought the introduction of Jeff Id's website to CA readers, and McIntyre wrote a comical post mocking James Hansen, who in an interview said that he would not use McIntyre's name, of course using McIntyre's name in the process. This was perhaps the nadir of the Michael Mann-inspired media strategy of ignoring critics. But the critics were growing in numbers and with the tool of FOIA, McIntyre demanded data from ex CRU employee Ben Santer, now working for Lawrence Livermore Labs in California.

Santer, evidently a man with a temper, fought against releasing the data but his employers said otherwise. Santer writes:

> *After reading Steven McIntyre's discussion of our paper on climateaudit.com (and reading about my failure to provide McIntyre with the data he requested), an official at DOE headquarters has written to Cherry Murray at LLNL, claiming that my behavior is bringing LLNL's good name into disrepute. Cherry is the Principal Associate Director for Science and Technology at LLNL, and reports to LLNL's Director (George Miller).*

In the end Santer will relent and post the data other researchers need to replicate his work. As 2008 comes to a close, Jones gives Santer the inside story of how he plans to handle CA requests for information. It's advice to Santer on what tactics work. Compromise the FOIA office by getting them to focus on the motives of the people making requests. Jones has worked the FOIA officers and convinced them that CA and its associates should be cut off. He has also deleted mails to thwart any FOIA.

From: Phil Jones <p.jones@uea.ac.uk>

To: santer1@llnl.gov, Tom Wigley <wigley@ucar.edu>

Subject: Re: Schles suggestion

Date: Wed Dec3 13:57:09 2008

Ben,

> *When the FOI requests began here, the FOI person said we had to abide by the requests. It took a couple of half hour sessions - one at a screen, to convince them otherwise showing them what CA was all about. Once they became aware of the types of people we were dealing with, everyone at UEA (in the registry and in the Environmental Sciences school - the head of school and a few others) became very supportive. I've got to know the FOI person quite well and the Chief Librarian - who deals with appeals. The VC is also aware of what is going on - at least for one of the requests, but probably doesn't know the number we're dealing with. We are in double figures. One issue is that these requests aren't that widely known within the School. So I don't know who else at UEA may be getting them.*
>
> *CRU is moving up the ladder of requests at UEA though - we're way behind computing though. We're away of requests going to others in the UK - MOHC, Reading, DEFRA and Imperial College. So spelling out all the detail to the LLNL management should be*

the first thing you do. I hope that Dave is being supportive at PCMDI. The inadvertent email I sent last month has led to a Data Protection Act request sent by a certain Canadian, saying that the email maligned his scientific credibility with his peers! If he pays 10 pounds (which he hasn't yet) I am supposed to go through my emails and he can get anything I've written about him. About 2 months ago I deleted loads of emails, so have very little - if anything at all. This legislation is different from the FOI - it is supposed to be used to find put why you might have a poor credit rating ! In response to FOI and EIR requests, we've put up some data - mainly paleo data. Each request generally leads to more - to explain what we've put up. Every time, so far, that hasn't led to anything being added - instead just statements saying read what is in the papers and what is on the web site!

Tim Osborn sent one such response (via the FOI person) earlier this week. We've never sent programs, any codes and manuals. In the UK, the Research Assessment Exercise results will be out in 2 weeks time. These are expensive to produce and take too much time, so from next year we'll be moving onto a metric based system. The metrics will be # and amounts of grants, papers and citations etc. I did flippantly suggest that the # of FOI requests you get should be another. When you look at CA, they only look papers from a handful of people. They will start on another coming out in The Holocene early next year. Gavin and Mike are on this with loads of others. I've told both exactly what will appear on CA once they get access to it!

Cheers, Phil

A short while after saying that he has deleted mails, Jones writes that he's been advised against deleting mails

From: Phil Jones <p.jones@uea.ac.uk>

To: santer1@llnl.gov

Subject: Re: A quick question

Date: Wed Dec 10 10:14:10 2008

Ben, Haven't got a reply from the FOI person here at UEA. So I'm not entirely confident the numbers are correct. One way of checking would be to look on CA, but I'm not doing that. I did get an email from the FOI person here early yesterday to tell me I shouldn't be deleting emails - unless this was 'normal' deleting to keep emails manageable! McIntyre hasn't paid his £10, so nothing looks likely to happen re his Data Protection Act email.

2008 ends with a promise near and dear to the hearts of CA readers. For as long as most of them have been reading the blog, McIntyre has been asking for a unique set of data used by Keith Briffa in his climate reconstructions: The Yamal dataset.

The Yamal dataset is a proxy temperature reconstruction using tree rings from Northern Russia—Yamal, which means The End of the World in the local language of the indigenous tribe that lives there. McIntyre had long suspected there were irregularities in the reconstruction, and had written about it on Climate Audit repeatedly.

On December 30th, 2008 McIntyre announces that Philosophical Transactions B of The Royal Society, the oldest scientific journal in the world and publisher of a Briffa paper, will enforce their data sharing rules and force Briffa to share his data. 10 months later, the records of 10 or so tree ring cores will end up in McIntyre's hands, and one year later Jones and Mann will be under investigation by their institutions.

It should be obvious that there is now open enmity between the two camps. The Team feels harassed and threatened—McIntyre has been correct in most of his assertions about their work. McIntyre and his band of irregulars feel that The Team would not be fighting so hard against the release of their data if there wasn't something to hide, and some of their tricks to hold on to the data are transparently obfuscatory.

McIntyre could only get data now by using FOIA requests, and usually only after appealing a first denial. This had the unexpected effect of giving The Team another weapon to use in their criticisms of McIntyre—that a barrage of FOIA requests were interfering with their work.

2009

In 2009 the FOIA battle for CRU data intensifies and the rationale for rejecting the requests gets more and more threadbare. At one point the MET posts a directive to readers that they can acquire raw temperature data from CRU. McIntyre contacted Dr. Kennedy of the MET and requested the data without citing FOIA. They directed him to Phil Jones. May 11th McIntyre promptly issues an FOIA to the MET who publish Jones' temperature index

I request the "archive of raw land surface temperature observations used to create CRUTEM3" as held by the Hadley Center (referred to on your webpage http://hadobs.metoffice.com/indicators/index.html) under the FOI Act or other applicable legislation.

The MET denied the request and argued that they did not hold the raw data, which is at odds with what they had previously told McIntyre. Previously they had held that they could not re distribute the data. On June 25th McIntyre makes another attempt, this time at CRU. This attempt is based on McIntyre's knowledge that Jones has sent the data to Peter Webster, a researcher. If the data is protected by confidentiality agreements, and it was sent to Webster, perhaps McIntyre can get it.

>Dear Mr Palmer,
>
>**Pursuant to the Environmental Information Regulations, I hereby request a copy of any digital version of the CRUTEM station data set that has been sent from CRU to Peter Webster and/or any other person at Georgia Tech between January 1, 2007 and Jun 25, 2009.**

Thank you for your attention, Stephen McIntyre

Over the course of the quest for that data McIntyre and his readers are given a number of differing answers for MET and CRU refusals, ranging from 'the data is already available', to 'the data can only be released to academics', to 'the agreements covering the data release have been lost', to 'the raw data itself has been lost'. Palmer's and Jones' plans were not very well coordinated.

McIntyre summed up the situation for readers on Climate Audit:

CRU Refuses Data Once Again

> Let me review the request situation for readers. There are two institutions involved in the present round of FOI/EIR requests: CRU and the Met Office. Phil Jones of CRU collects station data and sends his "value added" version to the Met Office, who publish the HadCRU combined land-and-ocean index and also distribute the CRUTEM series online.
>
> I requested a copy of the "value added" version from the Met Office (marion.archer at metoffice.uk.gov) which has been refused for excuses provided in my last post. On June 25, 2009, learning that Phil Jones had sent a copy of the station data to Peter Webster of Georgia Tech, I sent a new FOI request to CRU (david.palmer at uea.ac.uk) requesting the data in the form sent to Peter Webster. This too was refused today.
>
> We now have a new excuse to add to our collection of excuses – each excuse seemingly more ridiculous than the previous one.
>
> Dear Mr McIntyre
>
> ENVIRONMENTAL INFORMATION REGULATIONS 2004 – INFORMATION REQUEST (FOI_09-44; EIR_09-03)
>
> Your request for information received on 26 June 2009 has now been considered and it is, unfortunately, not possible to meet all of your request.
>
> In accordance with Regulation 14 of the Environmental Information Regulations 2004 this letter acts as a partial Refusal Notice, and I am not obliged to supply this information and the reasons for exemption are as stated below:
>
> Regulation 12(5)(f) applies because the information requested was received by the University on terms that prevent further transmission to non-academics
>
> Exception
>
> Reg. 12(5)(f) – Adverse effect on the person providing information   Information is covered by a confidentiality agreement
>
> Reason

> Regulation 12(1)(b) mandates that we consider the public interest in any decision to release or refuse information under Regulation 12(4). In this case, we feel that there is a strong public interest in upholding contract terms governing the use of received information. To not do so would be to potentially risk the loss of access to such data in future…

This is perhaps the most embarrassing excuse Palmer and Jones ever came up with. Jones has hinted at the existence of confidentiality agreements as far back as his emails to Warwick Hughes. But now he has been caught sending this data to Peter Webster. So now he claims that the confidentiality agreements preclude sending the data to "non academics." This would "explain" how Jones could send the data to Webster and deny the data to McIntyre. But McIntyre has three responses. As he and many of his readers point out, 'restriction clauses' usually target a "use" for example "non academic use." They do this because people who get access to data change jobs and because they can't check on the job status of people. It's also the "use" that people care about. If a third party sells the data, they will want to preclude "commercial use." The other response McIntyre has is that he is an academic. He is a published author in climate science and an expert reviewer of AR4. And the last response is the obvious. McIntyre has his coauthor McKitrick ask for the data. McKitrick teaches at a University. And finally McIntyre points out some inconsistencies between CRU and MET

> …over at the Met Office, they say "it cannot be determined which countries or stations data were given in confidence as records were not kept." But over at CRU, they purport to "know" nuanced details of the contractual language of the confidentiality agreements – clauses that have the effect of justifying the refusal of the data.

Taking the hint from McIntyre, several academics requested the data. If Palmer and Jones were telling the truth and the agreements allowed distributing to academics, like Peter Webster, then McIntyre's coauthor Professor McKitrick could request the data. That convenient invention of Palmers and Jones, would eventually turn out to be "a mistake" Webster was given the data by "mistake" and the agreements really precluding sending the data to anyone, or so the story of the lost agreements went. With this confusing description of the agreements on the table the next move was obvious: request the agreements. What did they actually say? Climate Audit readers signed up and each wrote an FOIA request for agreements if any existed. Readers picked countries and sent in their requests. As a result, CRU was forced to post up the agreements they still held. More bad record keeping. McIntyre reviews the documents and posts the following.

From Climate Audit:

CRU Excuses

> *Below we list the agreements that we still hold. We know that there were others, but cannot locate them, possibly as we've moved offices several times during the 1980s.*

Nobody would take it seriously. Nobody would believe that they were that incompetent. I wonder what would happen if Lonnie Thompson moved offices. Would he lose all his unarchived ice core data?

Or this excuse as to why they can't get data that any one of us can locate on the internet:

Much climate data are now additionally available through the internet from NMSs, but these are often difficult to use as data series often refer to national numbering systems, which must be related back to WMO Station Identifiers.

McIntyre continues:

But their trials and tribulations get even worse. They report that:

a number of NMSs make homogenized data ... available in delayed mode over the internet. Some that provide both raw and homogenized versions, generally do not link the two sets of data together.

Y'mean, that someone somewhere would actually have to inquire as to how to do the links. Hey, Phil, I've got an idea. If they won't tell you, send them an FOI. Or better yet, we'll save you some trouble. Make a list of all the NMSs that are troubling you and we'll send FOIs for you. Just have your people contact our people.

And as to why they haven't documented the source of their data. It's not their fault – that's impossible. The reason why they didn't document anything is that they "never had sufficient resources".

We are not in a position to supply data for a particular country not covered by the example agreements referred to earlier, as we have never had sufficient resources to keep track of the exact source of each individual monthly value.

At this stage it is important to recall Trenberth's complaint that the readers at Climate Audit did not understand how hard it was to maintain such a large database as a few thousand weather stations. That complaint seems more validly applied to CRU. McIntyre continues:

CRU refused my FOI request for CRU data stating that:

Regulation 12(5)(f) applies because the information requested was received by the University on terms that prevent further transmission to non-academics

I asked to see the precise language of the underlying agreements because I very much doubted that agreements specifically prohibited "further transmission to non-academics". If there were such a term in an agreement, it seemed far more likely that the term would be for "academic use" or something like that, and, given that my interest was scholarly rather than commercial, I doubted that the language of any applicable agreement would be applicable.

CRU has only managed to locate three documents pertaining to their agreements with NMSs: an application to Spain and letters from Norway and Bahrain, all from

1993-4. They also include a letter from CRU to the Met Office and, inexplicably, a copy of a current webpage from NERC governing Met Office data.

In their discussion of data availability, they state (in contorted logic) that:

In some of the examples given, it can be clearly seen that our requests for data from NMSs have always stated that we would not make the data available to third parties. We included such statements as standard from the 1980s, as that is what many NMSs requested.

Let's examine the agreements. If the effect of these and other agreements is to prohibit them from supplying me the data, it is under language that has far more broad reaching consequences than merely preventing the delivery of data to me.

The British Territories Agreement ... is the language from the CRU request to the UK Met Office regarding information about British Territories. It asks for data in connection with the construction of climatological normals; it makes no mention of the construction of a gridded temperature index nor the construction of a merged land-sea index. It says that the data would be used "unauthorized for any project" – which would obviously include its use in CRUTEM and HadCRU unless such authorization had been obtained. It doesn't prohibit delivery of the data to "non-academics"; it prevents delivery of the data to "third parties" -which includes all the recipients of Advance 10K data, those people who downloaded cruwlda2 or newcrustnsall and any other academic.

If this language is representative of agreements with NMSs, then it prohibits the delivery of the data to CDIAC at the US Department of Energy (who published versions in the mid1980s and placed a version online in the early 1990s.) For that matter, it would prohibit the delivery of NMS data to the Met Office for use in HadCRU as the Met Office is a "third party" to CRU.

To state the obvious, as far back as 2002 Jones shows an awareness of the existence of confidentiality agreements. The existence of those did not prevent him from sending the data to other researchers sympathetic to his positions, namely Webster and Mann. But, when the requester is a researcher who intends to "check" Jones' work, the agreements become an impediment. McIntyre continues with a recap of the other known agreements:

The Norway Correspondence The Norway correspondence is also from 1993-1994 (pre [WMO] Resolution 40) and also is about 1961-1990 climatological normals, rather than temperature data. But viewing this as an example of the "lost" agreements, as CRU invites us to do, it says that CRU must not give the data "to a third party". It doesn't say "non-academic" – it says "third party." If CRU seriously holds that this clause is in effect, then, as above, it prevents them from delivering data to the Advance-10K academics, to CDIAC and to the UK Met Office for use in HadCRU.

The Bahrain Correspondence The language may well not constitute a binding agreement. But let's stipulate that it does. Again it doesn't refer to "non-academics"; it refers to "third parties", including academics, CDIAC and the Met Office.

The Spanish Correspondence As others have noted, the Spanish correspondence contains no prohibition on delivery to third parties – it merely asks for citation. I suppose that CRU might argue that there was an implied term of the agreement regarding further transmission of the data – but on the available evidence, such implied term would extend to all "third parties" and not merely to Climate Audit readers.

As the FOIA controversy lay dormant, Briffa's Yamal tree ring data is posted up at the end of September, months after the journal announced that it would make Briffa comply with its policy of sharing data. McIntyre has years invested in pursuing this data. and he proceeds to fill the pages of Climate Audit in October 2009 and early November with a series of posts about the data that has been held back from him for all these years. Yamal dominated the Climate blogosphere and the story hit some of the mainstream media as McIntyre dissected the data.

CHAPTER SEVEN: HELL WEEK

Summary: Chapter 7 is a day by day account of what happened between the last Climategate mail being written on Nov 12th and the eventual discovery of the files on the web Nov 19. We see that whoever released the files had to work pretty hard to get someone to read them, trying Real Climate and Climate Audit before finally getting them posted on The Air Vent. One of your co-authors, Steve Mosher, actually played a part in getting the files disseminated, but has to confess to seeing a really bad movie to bring you the whole story. We speculate a bit on who released the files—the most parsimonious explanation being that the file had been collected during analysis of one of Steve McIntyre's FOIA requests, and was released by someone within CRU following the rejection of McIntyre's request.

Press Release: Dateline November 12, 2008.

> **UEA succeeds in Quest for secure IT access**
>
> **Published:12-November-2008**
>
> **By Steve Evans**
>
> **The University of East Anglia (UEA) has implemented Quest Authentication Services to enable single sign-on authentication for over 40,000 user accounts. The new system has dramatically reduced the number of help desk and support calls. The university has a number of IT systems, including desktop and email systems and a new portal, to which students are assigned a username and password. The authentication team at UEA is responsible for providing each user with secure access, but the university's old system did not enable automatic password synchronisation.**

One year later, on November 12, 2009

On Thursday Nov 12, 2009 at 10:18AM, Phil Jones sent his last mail in the Climategate files, the penultimate mail in the files. Apparently unaware that his and others' emails and documents were being harvested by someone, either a whistleblower inside CRU or a hacker who had broken into the system the University had upgraded just one year prior, Jones announces that he will be leaving work on Friday the 13th at noon.

> *From: Phil Jones p.jones@xxxxxxxx.xxx*
>
> *To: Sandy Tudhope <sandy.tudhope@xxxxxxxx.xxx>*
>
> *Subject: Latest draft of WP1 Date: Thu Nov 12 10:18:54 2009*
>
> *Cc: "Wolff, Eric W" <ewwo@xxxxxxxx.xxx>, Rob Wilson <rjsw@xxxxxxxx.xxx>, "Bass, Catherine" <C.J.Bass@xxxxxxxx.xxx>, "Turney, Christian"*

<C.Turney@xxxxxxxx.xxx>, Rob Allan <rob.allan@xxxxxxxx.xxx>, Keith Briffa <k.briffa@xxxxxxxx.xxx>, "t.osborn@xxxxxxxx.xxx" t.osborn@xxxxxxxx.xxx

Dear All (especially Chris/Catherine), Here's the latest draft of WP1. All in the group have now commented and amended this. You should have the 3 supporting letters from Tree partners. Eric was contacting Eric Steig and Sandy (see below) is contacting 3 coral people. There is an issue about a Map. Rob W put one in his PhD page. This shows the corals. If we were to add the tree-ring sites we would mainly get a splodge of points in South America and NZ. Ice cores would just be over the AP and in the low-lat Andes. Issue is one of space. We already have 3pp fo this WP. Refs will reduce to about 0.5pp once we go to et al for 3 or more authors. A map would be useful for presentation to NERC, but is it essential for the submission? I'm away from tomorrow lunchtime for the weekend. Back in on Monday. Hope we'll be looking through more complete drafts next week!

Cheers

Phil

This mail, the second to last in the collection, is unremarkable. While other mails detail bad behavior on the part of scientists that looks rather embarrassing, sometimes unethical and perhaps occasionally even verges on the criminal, and certainly all too human, this mail is notable for its banality. Its connection to the issue of McIntyre's FOIA requests is tangential at best. What's notable is that it contains some key externalities: It's a mail from Jones to Rob Wilson and Keith Briffa., principal players in one of McIntyre's data quests. And it contains some key words that would surely interest McIntyre: "tree-ring", "corals," "ice core". But there is no smoking gun here. No evidence of any scientific misconduct or even boorish behavior. Which raises the question "why is this here?" and what does it tell us about the person who collected these files? Did they know what they were looking at or merely collecting emails from key players? Or were they collecting emails from key players that mentioned certain key words such as "tree-ring" and "corals?"

A second clue, perhaps, can be found in a mail from Rob Wilson to Keith Briffa. Again, we find a mail that has nothing to do with scientists hiding data, coercing editors, or thwarting the FOIA process. It's a simple request from one scientist to another:

From: "Rob Wilson" <rjwilson_dendro@xxxxxxxx.xxx>

To: <K.briffa@xxxxxxxx.xxx>

Subject: data - Quaternary Science Reviews 19 (2000) 87-105

Date: Thu, 21 Oct 2004 15:53:21 +0100

Reply-to: "Rob Wilson" <rjwilson_dendro@xxxxxxxx.xxx>

Hi Keith,

When would be a good time tomorrow (or next week) to phone you about the data you have available at your website from your QSR 2000 paper. I am particularly interesting in using the long chronologies from the Polar Urals (Yamal) and Tornetrask. This is for Gordon's and Rosanne's NH temp recon update, so I thought I should have a chat with you before using the data.

all the best

Rob

This mail reveals nothing untoward about the scientists involved. Why is it here? What exactly is the point about a 2004 email from Rob Wilson to Keith Briffa about Yamal data? Perhaps it is included merely because it contains the names of two key players, Briffa and Wilson. And perhaps it was included because it contained the key word "Yamal." A mail harvesting program targeting the key words "Briffa" "Wilson" and "Yamal" would catch such a mail in its net. But surely such a program would catch other inconsequential mails as well. And why fill a blockbuster file of mails with such an inconsequential fact? Wilson asking Briffa about Yamal data in 2004?

For McIntyre and his followers, "Yamal" data was one of the Holy Grails of climate reconstruction. (The data is from tree rings used for temperature reconstructions, and there are concerns that a) not enough trees were used for studies and b) that other trees nearby that are also in other studies don't produce the same temperature results.) McIntyre suspected that something was wrong with the data and had sought the raw data for years, only to be rebuffed by Briffa and the science journals that published Briffa's results. The journals refused to require Briffa to release his data. He was taken at his word and the journals even refused to execute their stated data sharing policies. McIntyre was stymied in his hunt for Briffa's data. Perhaps more importantly the Yamal results were a bone of contention between Briffa and McIntyre during the writing of the infamous IPCC Chapter 6 of AR4. McIntyre was assigned as a reviewer for that chapter and Briffa had editorial control, control over a chapter which was supposed to objectively review work Briffa had previously written.

In 2008, McIntyre's inability to get his hands on Briffa's data changed when Briffa published in Philosophical Transactions of the Royal Society. McIntyre sent in his customary request for the release of data and to its credit the journal lived up to its very strict data sharing rules and Briffa quietly posted his data on his website in early September 2009. There was no announcement that the data was posted, no notice given to McIntyre that his request had been honored. McIntyre stumbled upon it, much as he had stumbled upon files from other climate scientists. What followed, starting on September 26, 2009, was a flurry of posts from McIntyre all aimed at illustrating the problems with the Yamal data that had been hidden for all these years. As McIntyre plowed through the data and posted articles in rapid fire succession, Real Climate called Tim Osborne of CRU for reinforcements:

From: Gavin Schmidt <gschmidt@xxxxxxxxx.xxx>

To: Tim Osborn <t.osborn@xxxxxxxxx.xxx>

Subject: latest Date: 28 Sep 2009 17:59:04 -0400

Hi Tim, I know Keith is out of commission for a while (give him my regards when you see him), but someone needs to at least give some context to the latest McIntyre meme.
....

None of us at RC have any real idea what was done or why and so we are singularly unable to sensibly counter the flood of nonsense. Of course, most of the reaction is hugely overblown and mixed up but it would be helpful to have some kind of counterpoint to the main thrust. If you can point to someone else that could be helpful, please do!

Thanks

Gavin

The blog wars over Yamal heated up through October and into November. Briffa was sidelined with health issues, and as journalists called on CRU and Real Climate to explain, Gavin Schmidt was reduced to posting graphs from an anonymous commenter, which proved to be laughably wrong. (The poster has been identified as a NASA employee, as is Gavin Schmidt). McIntyre jokingly referred to the poster as "Gavin's Guru." Since the "Guru" had posted his code, McIntyre quickly found the errors in the "guru's" work. It was easy to spot as the "Guru" had simply tried to modify a program that McIntyre himself had posted. The Climate Audit practice of posting code and letting others see your work succeeded. McIntyre had posted his code, the "Guru" had misused it and McIntyre quickly spotted the error. Gavin Schmidt was not so diligent in his review of the Guru's work and the difference between the Real Climate approach to working with others and the more transparent ways of Climate Audit was evident. As Real Climate looked for someone to help them respond, Briffa co-author Melvin was identified as the only person qualified to comment on McIntyre's work. But co-worker Tim Osborne deemed him a loose cannon and he was not allowed to respond:

Apart from Keith, I think Tom Melvin here is the only person who could shed light on the McIntyre criticisms of Yamal. But he can be a rather loose cannon and shouldn't be directly contacted about this

Tempers escalated to a point where Gavin Schmidt thought that McIntyre had deliberately launched his investigation of Yamal data knowing that Briffa was in the hospital. Tim Osborne of CRU, however quickly point out Gavin's mistake in this private mail:

131

29/09/2009, Gavin Schmidt wrote:

The fact is that they launched an assault on Keith knowing full

well he isn't in a position to respond. This is despicable.

On Sep 29, 2009, at 9:50 AM, Tim Osborn <t.osborn@xxxxxxxx.xxx> wrote:

Gavin,

be careful here, I think it more likely that McIntyre only learned of Keith's absence after he started posting about Yamal and the real reason for the timing of all this is that we made the Yamal tree- core measurements available about 2-3 weeks ago (in fact Keith had thought they had been made available before he fell ill, and only realized in early September that they weren't -- and asked for that to be rectified).

Cheers

Tim

As the data analysis unfolded on CA, at WUWT and in the press which was covering McIntyre's work, McIntyre called out directly to one of the few Climate scientists who would visit his site and engage him, Rob Wilson. Wilson had also used the Yamal data, but only a processed version of it, what is termed a "chronology." McIntyre was after the raw measurements, as the statistical decisions one makes in creating a chronology depend upon understanding the raw data. The following exchange between Wilson and McIntyre is documented on Climate Audit:

Rob Wilson:

October 17th, 2009 at 2:52 am

  Dear Steve et al.

As a response to your recent posts and also to your private e-mails, here's my 10 cents towards the use of the Yamal and/or Polar Urals data.   I will admit that the data description (Figure 2 and Table 1) in D'Arrigo et al. (2006) is somewhat misleading. ……

I will not be troubling the journal with a corrigendum as it does not change the results of the paper at all.   Finally, I want to clarify that I never asked Keith Briffa for the raw Yamal data. The simple fact of the matter is, I have great respect for Keith and I saw no point at the time in asking for raw data when there was a published RCS chronology for that location.

> Rob
>
> Steve: Rob previously commented in Feb 2006 as follows "I would have preferred to have processed the Yamal data myself, but like you, was not able to acquire the raw data. "
>
> and by email [Rob Wrote]: "Keith would not give me his Yamal raw data, but said that the Yamal series was a robust RCS chronology. "

The reason for including that 2004 mail from Wilson to Briffa now seems a bit clearer. The issue of whether Rob Wilson had requested the raw Yamal data from Briffa in 2004 is a marginal issue at best. But to CA regulars Wilson's truthfulness was in question. In 2006 Wilson appears to have commented that Briffa would not give him the data and in Oct 2009 Wilson argues that he never asked Briffa for the **raw** data. Yet the Climategate files contain a Wilson mail to Briffa in 2004 discussing the fact that Wilson wants to talk with Briffa before he uses the data.

> *Hi Keith,*
>
> *When would be a good time tomorrow (or next week) to phone you about the data you have available at your website from your QSR 2000 paper. I am particularly interesting in using the long chronologies from the Polar Urals (Yamal) and Tornetrask. This is for Gordon's and Rosanne's NH temp recon update, so I thought I should have a chat with you before using the data.*
>
> *all the best*
>
> *Rob*

Speculative questions abound. Was this mail included to support Wilson's claim that he had not asked for the raw data but rather the chronologies? Is Briffa clearing his friend? Or is it simply caught up in the net of an email harvesting program? Instead of being collected by an avid reader of CA, were the mails simply collected by using filters that looked for keywords found on Climate Audit.? In fact, there are quite a few mails, like out of office replies, that contain nothing of consequence that appear to have been collected by a program looking for key players and key subjects.

And what of the **very first** mail in the stack, a mail from 1996. This mail has no connection to the FOIA controversy. At best it only indicates some interesting and potentially questionable money transactions between Briffa and the Russian scientists who collected the Yamal data. It seems highly unlikely that Briffa, as speculated by some on the internet, would include such a mail.

> *From: "Tatiana M. Dedkova" <tatm@xxxxxxxxx.xxx>*
>
> *To: K.Briffa@xxxxxxxxx.xxx*
>
> *Subject: schijatov Date: Thu, 7 Mar 96 09:41:07 +0500*

Dear Keith, March 6, 1996   I and Eugene received your E-mail of 04.03.1996. This day I talked over the telephone with Eugene and he asked me to send an answer from   both of us.   Thank you for the information concerning proposals to the   INCO/COPERNICUS. We agree with your strategy used and we hope   that this proposal will not be rejected. ………. We can send to you all raw measurements   which were used for developing these chronologies.   Of course, we are in need of additional money, especially for   collecting wood samples at high latitudes and in remote regions.   The cost of field works in these areas is increased many times   during the last some years. That is why it is important for us   to get money from additional sources, in particular from the ADVANCE   and INTAS ones. Also, it is important for us if you can transfer   the ADVANCE money on the personal accounts which we gave you earlier and the sum for one occasion transfer (for example, during one day)   will not be more than 10,000 USD. Only in this case we can avoid big taxes and use money for our work as much as possible. Please,   inform us what kind of documents and financial reports we must   represent you and your administration for these money. I and Eugene have a possibility to participate in the Cambridge meeting in July, but we need extra many and special invitations. If you do not have enough money to invite both of us, Eugene does not insist upon this visit. The best wishes to you and Phil.

Yours sincerely

Stepan Shiyatov

The message left by the anonymous poster and the name of the folder (FOIA2009) would seem to indicate that the person who collected the files was interested in seeing that climate data requested under FOIA be released to McIntyre. Yet Yamal data was **already** released to McIntyre in early September of 2009 and he was making news with his analysis, and the Yamal data had never been the object of any FOIA request. Yet the files are replete with mails about Yamal, mails that fill in details only of interest to the most avid readers of CA. Details of no real scientific consequence.

If the Yamal mails in the stack gives us a clue that the collector knew what he was looking for, the very last mail is a complete mystery. On Nov 12 2009 the last mail in the Climategate files was sent.

> *From: "Thorne, Peter (Climate Research)" <peter.thorne@xxxxxxxxx.xxx> To: "Phil Jones" <p.jones@xxxxxxxxx.xxx> Subject: Letter draft*
>
> *Date: Thu, 12 Nov 2009 14:17:44 -0000*
>
> *Phil, attached is a draft letter. We were keen to keep it as short,   sweet and uncomplicated as possible without skipping over important   details. Shorter, simpler, requests are more likely to get read and   acted upon was the specific advice from international relations.   --*
>
> *Peter Thorne, Climate Research scientist*

Met Office Hadley Centre, FitzRoy Road, Exeter, EX1 3PB. tel. +44 1392 XXXXXX fax. +44 1392 XXXXXX

1 http://www.hadobs.org   Attachment Converted: "c:eudoraattachPhil_letter_draft_091109.doc"

The contents of the mail are unremarkable. It's Thursday afternoon a little after 2 PM and Thorne, who has also worked on the CRU temperature series, sends Jones a mail asking him to review the draft of a letter. The contents of the letter are unknown as the attachment was not included in the document folder of the Climategate files. From the mail all we can determine is that Jones will be sending out a letter, a short simple request. And that the letter was reviewed by international relations department. Jones will be sending a letter to some countries. What will Jones be requesting? Is this the beginning of Jones effort to get agreements from other countries to release climate data? And so the final Climategate mail seems to end with a whimper rather than a bang. Until we understand what happened on Nov 13, 2009 and perhaps why the collection of mails stopped on November 12th.

Friday the 13th November 2009.

On Friday the 13th Phil Jones planned to leave the office at noon. But the CRU FOI office was still at work writing a letter on the FOI appeal they had received from Steve McIntyre. In late July of 2009 McIntyre had appealed CRU's decision to reject his request for Phil Jones temperature data. In September of 2009 McIntyre had sent a follow up letter to the FOI department and pointed out some inconsistencies in their claim that they had to protect confidentiality. But on the 13th CRU had closed the case, explaining perhaps why the harvesting of mails had stopped on the 12th

The case was closed, a decision had been made. Their letter was written on Nov 13th 2009. It was not sent until days later. The letter is as follows:

ENVIRONMENTAL INFORMATION REGULATIONS 2004 – INFORMATION REQUEST (Our ref: FOI_09-44; EIR_09-03)

Pursuant to Mr. Palmer's letter of 21 September 2009 to you regarding the handling of your appeal of 24 July to our response of the same date in regards your FOI request of 26 June 2009, I have undertaken a review of the contents of our file and have spoken with Mr. Palmer and other relevant staff involved in this matter. I apologise for the delay informing you of my decision but we were awaiting the 'further particulars' in relation to this matter that you mentioned in your email of 2 September. Having not received such particulars, I have decided to proceed, given the passage of time, with my decision in their absence.

As a result of this investigation, I am satisfied that our overall decision to not disclose the requested information is correct.

In response to your first point in your email of 24 July regarding the non transmission of data to non-academics, I have concluded that the reference to non-academics was in error and that the status of yourself, or any other requester, is irrelevant to the factors to consider regarding disclosure of the requested information.

Turning to the points you raised in your email of 2 September, you note that other earlier versions of this data are available on the US Department of Energy website and that Dr. Jones had sent an earlier version of the data to you and had mounted it on FTP server.

In regards the information provided to the US Department of Energy, my investigation has revealed that this was done in the early 1990s prior to the imposition of the restrictions now pertaining to the data pursuant to a contractual obligation at the time. Therefore, the analogy you are drawing does not apply to the data that is the subject of this request.

In regards your second point regarding the provision of the data to yourself, and the fact that the information was mounted & left on our FTP site & also provided to Georgia Tech without securing consent of the institutions that provided it, we would, upon reflection, consider this an action that we not choose to take again. However, having made errors in past does not, in our eyes, justify making the same errors again.

I note that in your email of 2 September, you state that your request was 'for the current version of the data set' but in your original request, you asked for the subset of data that was sent to Georgia Tech University. I would advise that the many of the same restrictions apply to the full CRUTEM dataset as apply to the subset sent to Georgia Tech, but this analysis and answer is based on your original request.

In regards the substance of the exception claimed under Reg. 12(5)(f), I would maintain the position taken to date. There are restrictions on the release of at least some of the data cited, and our opinion is that any release would be contrary to the agreements, and release would have an adverse effect on those organisations. DEFRA guidance notes that the Aarhus Convention, which contains the origins of the Directive on which the EIRs are based, protects information volunteered by a third party and requires their consent to disclose it. The purpose of the exception is to encourage the free flow of information from private persons or institutions in order to protect the environment where making it available to the public could inhibit that process. To provide information that has a restriction on further transmission on it would not only damage CRU's ability to secure such information in future, but would also harm the interests of the organisations providing the information, who clearly have an interest in restricting transmission of the information due to the very existence of the restrictions.

Regulation 12(11) requires that we provide as much requested information as is possible outside the coverage of any applicable exception. After consultation with Phil Jones and other relevant staff in regards the nature and composition of the requested dataset, I have concluded that the data is organised in such a way as to make it extremely difficult and time-consuming to segregate the data in the manner that you suggest and would indeed, in our view, amount to an unreasonable diversion of resources from the provision of services for which we, as an institution, are mandated. Further, we would maintain that where no such segregation has, or will occur, we should not release any of the data for fear of breaching such restrictions as do exist.

I would note that we are, however, proceeding with efforts with the international community to secure consent from national meteorological institutions for the release of the information that they provide us with, and it is fully our intention to publish such data where, and when, we have secured such consent. This is in line with guidance from DEFRA that suppliers of volunteered information should be encouraged to consent to release where appropriate, and where it is lacking, such consent can be sought later in response to a particular request or in order to proactively disseminate the information.

In regards our obligation to assess the public interest in applying these exceptions, I am of the opinion that the public interest balance is in favour of non-disclosure of the requested information. As noted above, the public interest in maintaining the flow of information from institutions to CRU, and maintaining good working relations with international organisations, outweighs, in this case, the interest in the release of the data.

We have contacted the Information Commissioners Office in regards this matter and their advice is that if you are still dissatisfied with this response, you should, at this time, exercise your right of appeal to the Information Commissioner.

Yours sincerely

Jonathan Colam-French

For current purposes the legal issues involved in the dispute over the information can be put aside. The critical points here are as follows. Jonathan Colan-French had consulted a **file**. Was this file the Climategate mails itself? Were the mails harvested by the FOI office itself in the course of doing its investigation and did the mails stop on Nov 12th because the investigation was over? While the vast majority of emails contain many files and documents that have nothing whatsoever to do with McIntyre's FOIA or his appeal, the date the collection stopped and the messages delivered by the person who harvested the file both indicate that person who released the files knew about the progress of the FOIA appeal. Also, As Colam-French indicates, CRU continues to work with other countries' meteorological organizations to get permission to release their data, Perhaps a reference to the draft letter that Thorne has asked Jones to review in the last mail on Nov 12th.

There is another possible clue in the rejection letter. Colam-French makes reference to an earlier incident in 2009. In the July 2009 timeframe as McIntyre was issuing his FOIA request for Jones' temperature data, McIntyre pointed out that Jones had inadvertently published some of the data McIntyre requested on the CRU open ftp site. Jones, repeating the mistake he had mockingly chided Mann for, had left the data he so jealously guarded in broad daylight on the CRU open ftp site. That incident, played out in July of 2009 much to the delight of CA readers in a serious of 4 articles on where McIntyre engaged his readers in a series of whodunit posts, the first on July 25, where he announces that he is in possession of a version of the data he has been denied.

137

A Mole

by Steve McIntyre on July 25th, 2009

OK, folks, guess what. I'm now in possession of a CRU version giving data for every station in their station list . In their refusal letter, the Met Office described adverse consequences of disclosing CRU station data, an event that apparently would let loose the Four Horsemen of the Apocalypse.

During the run up to this incident, McIntyre and his readers had issued an FOIA to Hadley MET. The MET, which publishes Jones' series, had refused McIntyre's request and argued in overblown terms that release of the CRU data would harm international relations: McIntyre continues:

> **The Met Office stated:**
>
> **Some of the information was provided to Professor Jones on the strict understanding by the data providers that this station data must not be publicly released. ...**
>
> **If any of this information were released, scientists could be reluctant to share information and participate in scientific projects with the public sector organisations based in the UK in future. It would also damage the trust that scientists have in those scientists who happen to be employed in the public sector and could show the Met Office ignored the confidentiality in which the data information was provided.**

They continued:

> **the effective conduct of international relations depends upon maintaining trust and confidence between states and international organisations. This relationship of trust allows for the free and frank exchange of information on the understanding that it will be treated in confidence. If the United Kingdom does not respect such confidences, its ability to protect and promote United Kingdom interests through international relations may be hampered.**
>
> **Competitors/ Collaborators could be damaged by the release of information which was given to us in confidence and this will detrimentally affect the ability of the Met Office (UK) to co-operate with meteorological organisations and governments of other countries. This could also provoke a negative reaction from scientist globally if their information which they have requested remains private is disclosed.**
>
> **...to release it without authority would seriously affect the relationship between the United Kingdom and other Countries and Institutions.**

Part of the appeal McIntyre has for his readers is the sarcastic way he interacts with self important government officials. In this case McIntyre had made a formal FOIA request that the MET, which was in receipt of the CRU station data, release that data to him so that he could conduct an audit of it. The MET refused, arguing that CRU had signed agreements with some foreign countries not to release the data, to do so would harm international relations. In reality, as McIntyre pointed out in these posts, a "mole" at CRU had made that data available to him. McIntyre's "whodunit" played out much to the delight of his readers, a few of whom knew the secret, and the MET and CRU became aware of their security breach: McIntyre writes July 28th:

> **Late yesterday (Eastern time), I learned that the Met Office/CRU had identified the mole. They are now aware that there has in fact been a breach of security. They have confirmed that I am in fact in possession of CRU temperature data, data so sensitive that, according to the UK Met Office, my being in possession of this data would, "damage the trust that scientists have in those scientists who happen to be employed in the public sector", interfere with the "effective conduct of international relations", "hamper the ability to protect and promote United Kingdom interests through international relations" and "seriously affect the relationship between the United Kingdom and other Countries and Institutions."**
>
> **Although they have confirmed the breach of security, neither the Met Office nor CRU have issued a statement warning the public of the newCRU_tar leak. Nor, it seems, have they notified the various parties to the alleged confidentiality agreements that there has been a breach in those confidentiality agreements, so that the opposite parties can take appropriate counter-measures to cope with the breach of security by UK institutions. Thus far, the only actions by either the Met Office or CRU appear to have been a concerted and prompt effort to cover up the breach of security by attempting to eradicate all traces of the mole's activities. My guess is that they will not make the slightest effort to discipline the mole.**
>
> **Nor have either the Met Office or CRU have contacted me asking me not to further disseminate the sensitive data or to destroy the data that I have in my possession.**
>
> **By not doing so, they are surely opening themselves up to further charges of negligence for the following reasons. Their stated position is that, as a "non-academic", my possession of the data would be wrongful (a position with which I do not agree, by the way). Now that they are aware that I am in possession of the data (and they are aware, don't kid yourselves), any prudent lawyer would advise them to immediately to notify me that I am not entitled to be in possession of the data and to ask/instruct me to destroy the data that I have in my possession and not to further disseminate the sensitive data. You send out that sort of letter even if you think that the letter is going to fall on deaf ears.**

In the end McIntyre revealed that the mole was in fact Phil Jones and that Jones himself had placed this data on an open FTP server. The irony here of course is that it was Phil Jones who warned Mann about leaving files on FTP sites. For CA readers the case was pretty clear. The MET argued that releasing the data would damage international relations, but in fact Jones had already released the data, first to Peter Webster and then on his FTP site with no damage whatsoever to international relations. The MET's argument about potential damage from the release of the information had already been tested. The data had already been released, with no dire consequences. As of this writing, CRU and MET are making the same argument.

After Jones was revealed as the leaker of his own data, the CRU directed employees to purge a whole host of information from their open FTP sites, including information that apparently shows up in the documents folder of the Climategate files. McIntyre ended his posts on the "Mole Hunt" with a piece on August 4th entitled "Dr. Phil Confidential Agent"

> **Recently, Philip Jones of CRU (Climatic Research Unit) claimed to have entered into a variety of confidentiality agreements with national meteorological services that prevent him from publicly archiving the land temperature data relied upon by IPCC.**
>
> **Unfortunately, Jones seems to have lost or destroyed the confidentiality agreements in question and, according to the Met Office, can't even remember who the confidentiality agreements were with. This doesn't seem to bother the Met Office or anyone in the climate "Community". This sits less well with Climate Audit readers, many of whom have made FOI requests for agreements between CRU and other countries throughout the world.**
>
> **Because Jones is having so much trouble remembering who he made confidentiality agreements, we here at Climate Audit, always eager to assist climate scientists, are happy to do what we can to help Jones' memory. I've spent some time reviewing Jones' publications on the construction of the CRU_Tar index – in particular, Jones et al (1985); Jones (1994); Jones and Moberg (2003) and Brohan et al (2006). These contain interesting and relevant information on the provenance of Jones' information and provide helpful clues on potential confidential agreements.**
>
> **Jones insists on the distinction between academics and "non-academics" being scrupulously observed. I honored Jones' demand that this distinction be observed by using his full academic title, Dr Phil, in the title, but, in the rest of the post, for the most part, I will refer to him more informally merely as Jones.**

As one reads through the Climategate files it's important to place the emotions that scientists display toward McIntyre in context. The venom they display toward him seems odd until one realizes that McIntyre is at once rigorously mathematical and brutally funny and sarcastic at their expense. But more than that he has no compunction whatsoever about demonstrating their apparently utter incompetence in matters like simple data management and document control. Worse, from their point of view, it's all happening in the public eye.

In this case, the MET and CRU have argued that they cannot release the data because Jones remembers that he signed agreements with a few countries that preclude him from releasing the data. But, unfortunately, those agreements have been misplaced, so they can't be reviewed, although Jones claims to recall (wrongly it turns out) that non academics like McIntyre are not allowed to see the data. Armed with the 2003 version of Jones' data left by Jones on the CRU FTP, copies of Jones's publications, and the open records at GHCN, McIntyre applies his formidable sleuthing skills and figures out a short list of countries that could possibly have agreements with CRU. He concludes his posting with the following:

> **If one is trying to narrow down the search for the lost CRU confidentiality agreements, it seems to me that the logical place to start is with the following nine countries: Iran, Algeria, Taiwan, Croatia, Israel, South Africa, Syria, Mali, Congo. Did Dr Phil enter into confidentiality agreements with all or some of these nine countries? If so, what are the terms? And why are these agreements "lost"?**

> And exactly when were these agreements entered into? Our analysis shows that these agreements are not hoary old agreements from early Oak Ridges-CDIAC days, but were made well within the IPCC period. As a long-time IPCC hand, Jones knew of the IPCC commitment to "openness and transparency".
>
> Did Jones seek approval from an advisory board or some other form of independent oversight prior to unilaterally entering into confidentiality agreements with various totalitarian (and other) regimes? If these few confidentiality agreements are sufficient to poison public disclosure of the entire database (as CRU and the Met Office now argue), how was Phil Jones able to poison the public availability of the database through a few seemingly unsupervised confidentiality agreements?

The Mole scenario illustrates the clash of various cultures at play here. On one hand we have the IPCC and government agencies such as the MET, governed by principles of openness and transparency and steeped in a tradition of document control. We also have a culture of auditors and engineers at CA, steeped in a tradition of due diligence and full documentation of every calculation and caught in the middle is a research scientist, whose scientific principles dictate openness but whose work habits preclude it.

Finally, the Mole escapade makes another possibility evident. While the timeline of the mails and McIntyre's FOIA rejection points to a whistleblower on the inside, perhaps an avid reader of CA, the security practices at CRU points to another possibility: CRU left the Climategate files in full view much the same way Jones had left the data McIntyre requested in full view.

November 14 – November 15th

Phil Jones was off for the weekend, but somebody somewhere was apparently at work, preparing a file that would eventually be uploaded on November 17th. The evidence that some sort of work was performed on the files in question is rather scant. There are two folders in the collection, one entitled mails and the other entitled "documents." Every mail document has had its creation date modified to Jan 1, 2009, 12AM, effectively hiding the time the documents were copied. In the documents folder the changing of document creation dates is haphazard with only selected documents with their dates changed to Jan 1 2009, 12AM. In addition some documents appear to have been backdated to 1980. What external hacker would want to hide the date that files were copied? The only person who would want to hide the dates that the files were copied would be an insider. For example, someone in the office on the weekend.

Monday November 16th.

Nothing remarkable happened on November 16th and McIntyre posts a short article mid day. The topic is somewhat arcane, related to a longstanding feud McIntyre has had with Michael Mann about the use of strip bark trees. Taking a hint from an infrequent but knowledgeable commenter at CA, McIntyre directs his readers to an article written by none other than Rob Wilson's director:

> A CA reader has provided a link to an extremely interesting presentation by dendro Brian Luckman of U of Western Ontario (Rob Wilson's thesis supervisor) at the 2008 Canadian Society of Petroleum Geologists. Reader Erasmus de Frigid draws attention to the inhomogeneity in the tree ring record created when the tree was scarred by a glacier, evidenced by a terrifically interesting cross-section picture of the results of glacier scarring on ring widths.

November 17 Tuesday, 620 AM EST: The Break-in at Real Climate

At dawn somebody accessed the Real Climate account and posted a copy of the Climategate files. In Norwich, the home of CRU, it was just before lunch. Gavin Schmidt posted the following timeline on Real Climate:

There seems to be some doubt about the timeline of events that led to the emails hack. For clarification and to save me going through this again, this is a summary of my knowledge of the topic. At around 6.20am (EST) Nov 17th, somebody hacked into the RC server from an IP address associated with a computer somewhere in Turkey, disabled access from the legitimate users, and uploaded a file FOIA.zip to our server. They then created a draft post that would have been posted announcing the data to the world

There seems no reason to doubt Schmidt's recounting of the timeline. Somebody accessed the RC server and created a post, its contents still withheld by RC, and uploaded a file FOIA.zip which contained the Climategate files. The access is described as a "hack" which implies an unauthorized access, but RC offer no evidence of this. If one scours the files that were released two things are clear: there was a stream of mails from RC to CRU, calls for help to fight the Yamal story that McIntyre was running, and on several occasions various passwords and login IDs were being exchanged via mail. Until RC claims otherwise the possibility exists and should be seriously entertained that somebody on RC sent a mail to CRU with a password to the RC server. The password would allow a CRU employee to upload a post to RC, perhaps defending Briffa's Yamal work. Thus a whistleblower inside CRU who was collecting files and emails would find the information he needed to post the incriminating files on the servers maintained by the very website that was set up to control the story of climate science, an irony not lost on the readers of CA.

The post, however, was never published. But the creator of the file had a backup plan for announcing the data. A plan that hid a link to the file in plain sight on Climate Audit. At 7:24 EST the following post appeared on Climate Audit as a comment on the post McIntyre had put up on Nov 16th. CA runs its site without moderation. That means readers can post whatever they like, unlike RC. This openness often leads to rude behavior, flame wars, and off topic comments much to the dismay of people who do not frequent the site on a daily basis. For the most part McIntyre will go through comments after they are posted and delete or snip the inappropriate comments. In the case of the link to the file on RC, McIntyre missed the comment altogether and never saw what was hidden in plain view:

142

Figure 13: Screen Capture of Climate Audit, November 17, 2009

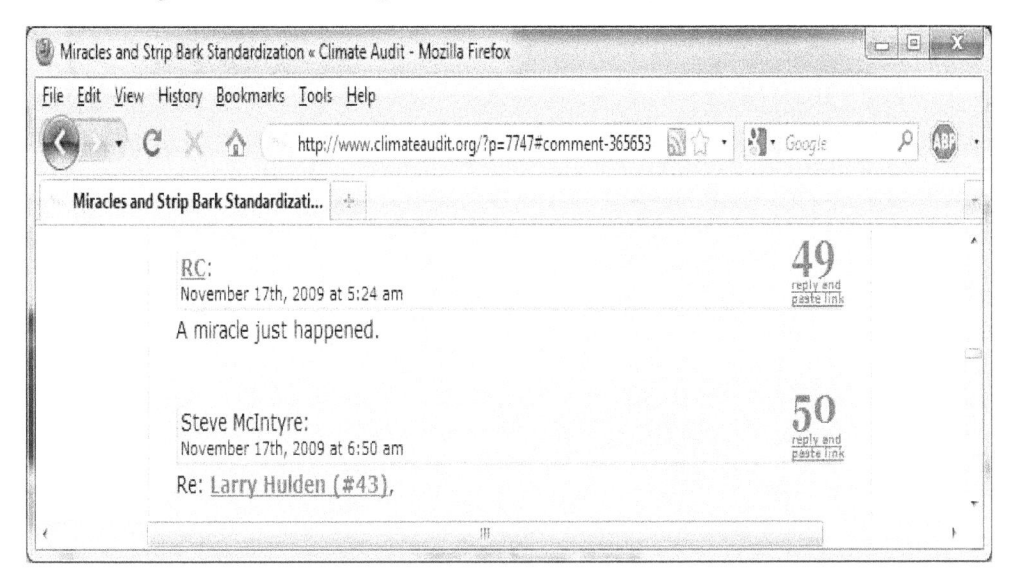

Reproduced by permission, Climate Audit

The comment, comment 49, was posted to the miracle thread and 1 hour and a half later McIntyre posts comment 50, oblivious to the comment above: "a miracle just happened" posted by a commenter who has taken the name "RC". But the name "RC" is more than just a name, it's a link. A link to the Climategate files hosted on Real Climate. Had McIntyre clicked on the link he would have found that it directed users to the zip file containing the Climategate files. As it transpired, McIntyre and his readers, all long time regulars, appear to have totally overlooked the comment as they were engaged in a discussion of strip bark trees. At 8:58 eastern time, a special guest appears to comment:

Rob Wilson:

November 17th, 2009 at 6:58 am

Dear All,

Please do not take this the wrong way, but the depressing amount of ignorance on this current thread makes it hardly worth the effort to respond. Two quick points: 1. The linear aggregate model is a purely conceptual model – something ideal for teaching to undergraduates to highlight all the environmental factors that can affect tree growth. 2. w.r.t. scarred trees (fire, glacier, avalanches etc) one would never use tree-ring data from a scarred tree for a dendroclimatic reconstruction – or at the very least one would use a measured radius where the rings were not affected by the accelerated growth around the scar.

There is a wealth of literature on these issues.

Rob

There is no suggestion that Rob Wilson knew about the link or what it linked to. But his appearance always increases traffic in the comments section. For hours to come the regulars at CA engaged each other in the typical back and forth banter about the article. None appears to have taken the time to click on the link "RC" and why should they, as more often than not these links merely provide the email address of the poster or link to some site advertising products. According to Schmidt, however, the file on RC was accessed at least 4 times before McIntyre removed the comment. As Gavin writes

Curiously, and unnoticed by anyone else so far, the first comment posted on this subject was not at the Air Vent, but actually at Climate Audit (comment 49 on a thread related to stripbark trees, dated Nov 17 5.24am (Central Time I think)). The username of the commenter was linked to the FOIA.zip file at realclimate.org. Four downloads occurred from that link while the file was still there (it no longer is).

Nov 17th 6:25 PM PST.

The file had now sat at Real Climate for the better part of the day and a comment had sat on CA linking to that file, but none of the regular netizens who discuss climate science on a daily basis had written a word about it. Had the collector of the files been too clever with his wry comment on Climate Audit? Only the most curious would click on the link and then download an unknown file. A bolder more visible revelation was perhaps in order. At 6:25 PM PST CTM, a moderator at Watt's Up With That sat down for his daily chore of approving comments on the blog. Like Real Climate WUWT is a moderated blog. Comments must be reviewed and approved for publication. What the moderator saw gave him pause: A comment in moderation that claimed to point to a file containing secret mails of climate science.

Figure 14: Screen Capture, Watt's Up With That, Nov. 17, 2009

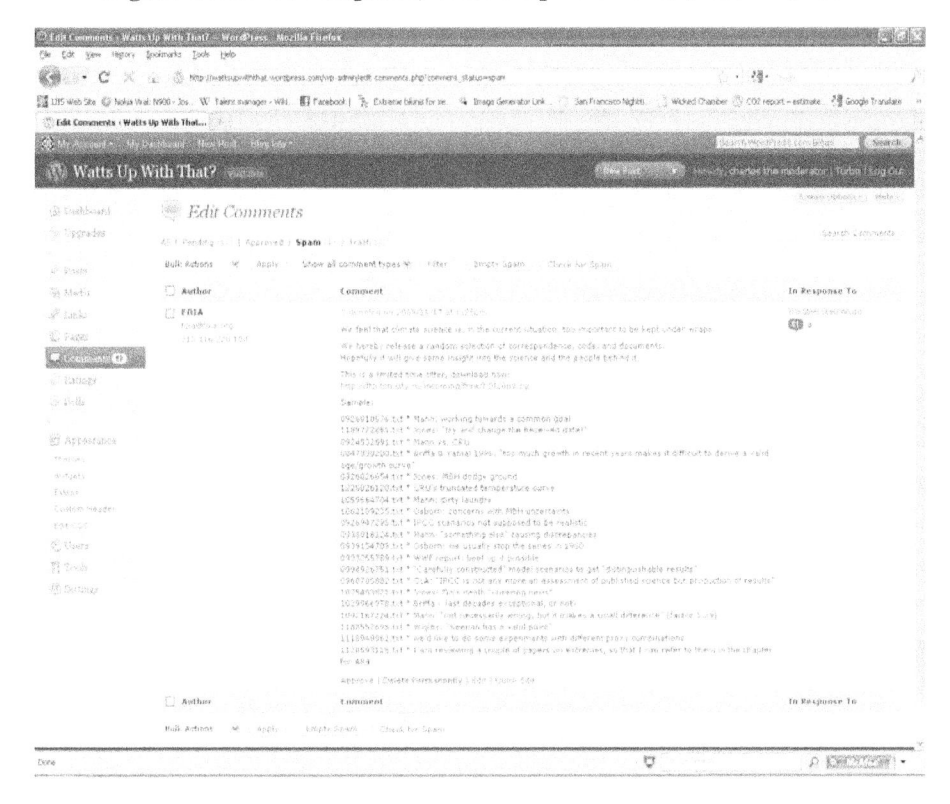

Reproduced by permission, Watt's Up With That

The first order of business with such a comment from an unrecognized poster is to make sure that the link does not transport visitors to any inappropriate content such as pornography. The moderator's first impression was that a link to a Russian server was highly suspicious, as his years in the computer industry had taught him; the site was an open proxy which meant whoever put it there had covered his tracks. The moderator suspected a cyber attack. He quarantined the comment and sent notes to all the other moderators to leave the comment alone while Anthony Watts was contacted. He downloaded the file, initiated a virus scan, and started an exchange of emails with Anthony Watts, who was out of the country.

Watts told us of his reaction to the discovery. ""I first saw the comment left on WUWT about 4 hours before I was to give a presentation at the EU parliament building in Brussels. The timing concerned me for two reasons. First, that I was sitting on a potential bombshell story but I didn't have much time left to study it all and determine its validity before I was to leave for that presentation. Secondly, since I didn't know at that time that the file link to the Russian FTP server had also been left on other blogs, including RealClimate earlier, and The Air Vent at about the same time as the WUWT comment, I was concerned that perhaps the file was some sort of a "setup" that would cause somebody to point a finger at me when I went through the EU security. I had been advised that the security process to get into the EU was an hour long affair and far more intensive than airport security. Thus for I opted on the side of caution and decided not to publicly announce the file until I had completed my presentation there and I had more time to study it, plus get some legal advice on the ramifications.

I decided it best to wait until I had returned to the US, so that I could seek expert advice, and so that if indeed the file was hacked/stolen, I would not be transporting it into the USA. I purchased and downloaded a DoD rated disk data wiper program to remove that file from my laptop. When I landed in Dulles the next day, my cell phone lit up with messages, and I soon discovered that The Air Vent comment noting the FOIA2009.zip file location had been seen by some people (even though it had been taken down) and the file was already making the rounds. I also learned that some people who had studied it, including McIntyre, believed it to be valid. Even though I opted for caution, the cat was out of the bag, so to speak.

I consulted with an attorney via telephone on the legal issue surrounding the FOIA2009.zip file, then I sat down at the gate in Dulles Airport, connected to WiFi, and wrote my first article on what would shortly be known as "Climategate", as coined by one of the commenters in that article. Literally, I finished that story with seconds to spare before the door to the airplane closed. While I was flying westward, my story was picked up by dozens of other web news outlets. By the time I landed in California 5 hours later, the WUWT story had gone viral."

At 7:02 PM CTM attempted to contact CA regular Steven Mosher, who missed the call. Mosher called back at 7:32 PM. The discussion was short and to the point, The moderator wanted to meet with Mosher to show him some files and get his opinion on their authenticity. At 9:36PM Mosher was handed a CD and he started reading. As he plowed through the mails, he recognized all the names and issues. The dates hung together with what he knew the timelines to be, mails about Yamal, about Santer and Douglas and when he found mails from McIntyre and one from his friend Lucia the next course of action seemed clear.

Before the call went out to McIntyre, however, some rules were laid down. The link was posted at WUWT. Anthony instructed CTM to do a screen grab and delete the comment. No one was to pass that link onto anyone else until Anthony was back in the country with all the facts laid out before him. A veteran of the news business, he knew that publishing a link to a hoax would devastate his reputation. With those rules in place Mosher read emails to McIntyre over the phone, first those that appeared to be from McIntyre himself and then others to corroborate the timelines. Mosher read late into the night while Real Climate alerted CRU to the potential breach in security.

Wednesday Nov 18th.

Throughout the day the reading continued and more calls were placed to McIntyre. Walking through the emails and files, it occurred to Mosher that they looked like a file that would be put together as part of an FOIA decision process. But the last mail was Nov 12th and Mosher's own FOI had been rejected long before that and he had decided not to appeal. Why would CRU FOI office continue to collect files? A quick call to McIntyre and the picture at CRU became clearer. McIntyre informed Mosher that he had just received a letter from CRU on Nov 18th, a rejection of his FOI appeal, dated Nov 13th. More circumstantial evidence that the files Mosher held were authentic and possibly a part of a FOIA investigation that had been closed on Nov 12th. Meanwhile at CRU, staff were informed of the breach and various security measures were put into place.

Thursday Nov 19th.

On the morning of November 19th word of the discovery of the breach at CRU reached Mosher. "I was told that CRU was informing employees of a breach in security and that a file had been posted somewhere on the net." If CRU was telling employees that this file was posted somewhere then it must be referenced somewhere else other than WUWT.

Apparently, the file link had been posted at sites other than WUWT, where it was embargoed. The link was out there, but where? The whistleblower or hacker wanted the link out, but it was being held by Watts and his team and ignored at CA. Without a link name to search for the best bet seemed to check the sites without moderation, looking for a needle in a haystack. Buried at Jeff Id's site (The Air Vent) Mosher found the following comment.

> **Foia said   November 17, 2009 at 9:57 pm We feel that climate science is, in the current situation, too important to be kept under wraps.   We hereby release a random selection of correspondence, code, and documents.   Hopefully it will give some insight into the science and the people behind it. This is a limited time offer, download now:**
>
> **http://ftp.tomcity.ru/incoming/free/FOI2009.zip**
>
> **Sample: 0926010576.txt * Mann: working towards a common goal**
>
> **1189722851.txt * Jones: "try and change the Received date!"**
>
> **0924532891.txt * Mann vs. CRU 0847838200.txt * Briffa & Yamal 1996: "too much growth in recent years makes it difficult to derive a valid age/growth curve"**
>
> **0926026654.txt * Jones: MBH dodgy ground**
>
> **1225026120.txt * CRU's truncated temperature curve**
>
> **1059664704.txt * Mann: dirty laundry**
>
> **1062189235.txt * Osborn: concerns with MBH uncertainty**
>
> **0926947295.txt * IPCC scenarios not supposed to be realistic**
>
> **0938018124.txt * Mann: "something else" causing discrepancies**
>
> **0939154709.txt * Osborn: we usually stop the series in 1960**
>
> **0933255789.txt * WWF report: beef up if possible**
>
> **0998926751.txt * "Carefully constructed" model scenarios to get "distinguishable results"**
>
> **0968705882.txt * CLA: "IPCC is not any more an assessment of published science but production of results"**
>
> **1075403821.txt * Jones: Daly death "cheering news"**

1029966978.txt * Briffa – last decades exceptional, or not?

1092167224.txt * Mann: "not necessarily wrong, but it makes a small difference" (factor 1.29)

1188557698.txt * Wigley: "Keenan has a valid point"

1118949061.txt * we'd like to do some experiments with different proxy combinations

1120593115.txt * I am reviewing a couple of papers on extremes, so that I can refer to them in the chapter for AR4

The picture seemed clear—whoever leaked this file was placing clues to it at WUWT where it was embargoed and on Jeff Id's Air Vent. Mosher informed CTM that the file was in the open, and that he had heard that CRU had verified that a file had been posted and that whatever agreement there was with Anthony to sit on the information had been overtaken by events. If the file was a hoax it was a very elaborate one. It was only a matter of time before somebody reading Jeff Ids site would come across the comments.

He sent a short email to Jeff Id just past 11:30, requesting a phone conversation and then went to Lucia's RankExploits and posted the following at 11:55 (1:55 in blogtime), assuming that Lucia would see the comment in the moderation queue and be alerted to the presence of her mail in the file.

Steven Mosher (Comment#23722) November 19th, 2009 at 1:55 pm

Lucia,

Found this on JeffIds site.

http://noconsensus.wordpress.c.....en-letter/

It contains over 1000 mails. IF TRUE ...

1 mail from you and the correspondence that follows.

After informing Lucia and her readers of the existence of the file containing an email belonging to her, Mosher alerted Andrew Revkin of the New York Times using Facebook at 12:14 PM PST. Mosher says, "My sense was that I needed to inform the people I knew first. McIntyre and I had had dinner in San Francisco back in 2007 and he could help verify those mails, Lucia has always been a favorite of mine and benders (a CA regular); I won a bet on her site and she baked me brownies. Revkin, I saw his name in one of the mails. I comment on his Facebook page regularly so I figured he deserved a heads up. Then I went to watch the movie 2012 and when the chief scientist walked into the President's office and said "I was wrong" I wondered if anyone at CRU had ever said those words"

Within a few minutes Lucia had a post up

> **Real files or fake?**
>
> **19 November 2009 (14:09) | politics Written by: Lucia**
>
> **Steve Mosher alerted us to an interesting development: Someone dropped a link to a zipped directory of files that contain what appear to be emails between various bloggers and climate science illuminati Of course this may be some sort of scam. If so, someone spent a lot of time putting together fake email/code etc.**

After the post on Lucia, the story spread quickly. A post on WUWT attracted huge readership, and traffic tripled. Climate Audit was stopped dead in its tracks by the huge load and a mirror site was set up to handle visitors. Five hours after Mosher sent his message to the NYT, Revkin responded that he would look into it, but by that time the netizens were well on to the case.

One story that played very well on the internet was the "whodunit." Looking through the documents it is easy to see patterns that suggest the file was built with care by someone familiar with all the issues CA was interested in. Enticing as that might seem, evidence in the file suggests otherwise. If a human being collated the files, then there are numerous inclusions, housekeeping mails that make no sense. They are noise and suggest that the file was created or harvested by automatic means: One such mail:

> *From: Eric.Steig@sas.upenn.edu (via the vacation program)*
>
> *To: k.briffa@uea.ac.uk*
>
> *Subject: away from my mail*
>
> *Date: Fri, 10 Nov 2000 09:53:09 -0500 (EST)*
>
> *I am away for a couple of days. This is an automatic reply. I will reply*
>
> *to your mail regarding "reminder" when I return on Sunday.*

There are numerous other examples of this type of mail which suggest that whoever collected the mails used a filter of sorts to collect the mails and documents. Such a filter would look for key names and or key words. There is also the possibility that the files were collected as a part of CRU internal FOIA review process. The coincidence of the McIntyre's FOIA appeal being rejected on the Nov. 13th and the last mail being sent on Nov 12th is highly suggestive that the two events have something to do with each other. But there seems to have been some editing, as well. There are no strictly personal emails, no 'honey I'll be late for dinner' messages of the type a harvesting program would yield.

Across the internet speculation has run from those arguing that Russian oil interests had something to do with the hack, to a suggestion that Keith Briffa himself was the whistleblower. As the mails suggest, Briffa was growing tired of all the attention, tired of Mann, and was under enormous personal pressure. Also, one System Administrator did a forensic analysis of the mails and "determined" that an external hack was extremely unlikely. At the time of this writing CRU continues their investigation, aided by the police.

The fascination with "whodunit" however illustrates a problem at the very heart of Climategate: the concern over motives. On the assumption that CRU was hacked by "evil forces," defenders of climate science are quick to dismiss what was actually found. But whoever collected the files did not change what was said. If CRU was hacked the hacker did not put words in Jones mouth. The argument that Climategate is some kind of attempt at character assassination backfires: Jones committed character suicide. If CRU was hacked, the hacker merely displayed the death scene. Gruesome, but nothing in his motives changes the facts of the files. On the other side we have the whistleblower hypothesis. This is also equally irrelevant. It matters little if the creator was Briffa or anyone else inside CRU. The facts of the files are what should draw people's attention.

The allure of determining motives, it should be recalled, is what precipitated the whole Climategate fiasco. Up until 2005 Jones appeared willing to share data with Warwick Hughes. In 2002 he promised to send files to McIntyre. Yet with the publication of MM03, Jones changed. And he changed most probably because of the influence of Michael Mann, who saw bad motives behind every request for data. That led to Jones' denial of Hughes' request for information on Feb 21, 2005, where Jones argued against releasing data to Hughes because Hughes' "aim" or motive is to find mistakes. But this is exactly why data is shared. Data is shared so that one researcher's potential errors, potential bias, potential motive can be checked by another researcher.

When that request was denied, people began to question Jones' motives; and his pattern of avoiding requests over the next 4 years only heightened the suspicion.

CHAPTER EIGHT: THE CRUTAPE LETTERS

Nothing in the Climategate mails contradicts the science. This is, of course, trivially true. The emails are not science. The best the Climategate files can do is give us insight into the process of climate science, that is, it can give us insight into the individual scientists and institutions that shape climate science. In a conversation with Stephan McIntyre he likened the behavior he saw to "noble cause corruption."

Noble cause corruption in policing is defined as "corruption committed in the name of good ends, corruption that happens when police officers care too much about their work. It is corruption committed in order to get the bad guys off the streets… the corruption of police power, when officers do bad things because they believe that the outcomes will be good." (Crank & Caldero) Examples of noble cause corruption are, planting or fabricating evidence, lying on reports or in court, and generally abusing police authority to make a charge stick.

Police officers tend to see bending of the rules for the greater good as acceptable rather than as misconduct or as corruption. They rationalize such behavior as part of the job they were paid to do and are what the public wants. It is seen by some in a utilitarian sense, to get the criminals off the streets, regardless of the means employed. Interested readers can find more on this subject at (http://www.ethicsinpolicing.com/noble-cause-corruption.asp)

Naomi Oreskes, a historian, is often referenced by believers in Global Warming. Her views inform the position of scientists like Mann. In her view global climate skeptics are employing the tactics that big tobacco employed in their fight against the science that showed the harms from smoking. The scientists also viewed the skeptics through the lens of David Michael's book, "Doubt is their Product," which detailed how the tobacco companies worked to create doubt in people's minds about the hazards of smoking. On this view climate skeptics are discounted because they are backed by corporate interests whose motivation is the preservation of profits. Moreover, the main tactic they feel that the skeptics use, like the tobacco companies, is uncertainty and doubt. The climate skeptics, backed by financial interests, publish questionable science in questionable journals to raise doubt. And doubt is important to skeptics because they want to delay action. And so there are three components that these critics of skeptics always look for: 1) the financial, political and personal interest of the skeptics, their ulterior motive 2. The publication in compromised journals and, 3. The strategy of raising doubt with the end of delaying action.

We hold that what the Climategate scandal shows is that The Team were captured by the worldview they thought they were fighting—that to a large extent they adopted the strategy and tactics that they ascribed to their opponents in this struggle.

Whether they are justified in adopting the tactics of their opposition may depend in part on how accurate their characterization of skeptics is. Are skeptics well-financed, well-organised, and intent on sowing doubt in the minds of the general public? Is there an organized skeptic front that needs an equally organized response from environmental activists like The Team and their supporters?

There are organizations, ranging from The Heartland Institute to the Competitive Enterprise Institute that have a consistent agenda of skeptical opposition to global warming, and in the past they have received funding from energy companies—but as we try to show in Chapter 9, funding of skeptics has not been either lavish or continuous, and most of these organizations are really more independent or Republican think tanks that oppose almost all liberal initiatives, including global warming. This has made them rather predicable, slow-moving and fairly easy to counter, and we certainly don't think they 'drive' the skeptic agenda. Indeed, it seems very clear to us that Anthony Watts and Steve McIntyre are far more leaders than followers, and as the two most active voices on the internet, they seem far more influential (and crucially, far less skeptical on the actual existence of anthropogenic-caused climate change) than the Heartland Institute, CEI, American Thinker and other skeptical organizations and publications.

The case against the skeptics was as follows: they were guided by ulterior motives, they published in suspect journals and they tried to create doubt. What we show is this. The Climategate scientists sold certainty where there was doubt, by manipulating the process of publication, because they had interests to protect—personal and financial interests.

Hide the decline

There are two issues that seem to have informed the debate, the FOIA requests, and the mails themselves: The issue of the surface record and climate reconstructions. The scientists fought to prevent the release of the data of the surface records and the code which produces the final figures. The discussions over the surface record can be distilled down into three basic positions. Those who follow Jones would argue that the land record (or 30% of the globe) shows very little influence from UHI. At the other extreme skeptics typically argue that a large portion of the temperature rise in the land record, on the order of 50%, is the result of UHI. Even the skeptical account indicates the globe is warming. In between these two positions is the Lukewarmer position. Rather than opening up the data and the processing to others, Jones fought from 2002 to the present day to keep the inner workings of his science secret. Rather than risk the possibility of someone showing that his estimate of .05C of UHI warming was off, he risked the whole mission of CRU. The uncertainty he created by not addressing the problem in a forthright manner is greater than the uncertainty in the underlying question. CRU has now decided to rework the entire index from square one. If they adopt an open approach they have a chance of not inflicting anymore damage on their credibility. Already, however, questions have been raised about CRU's treatment of Russian data.

The second issue, climate reconstructions, overlaps the first in an interesting way and forms the basis of the most troublesome scientific issue in the Climategate mails. The issue has famously been referred to as "hide the decline," taking a phrase from one of Phil Jones' mails:

> *From: Phil Jones <p.jones@xxxxxxxxx.xxx>*
>
> *To: ray bradley <rbradley@xxxxxxxxx.xxx>,mann@xxxxxxxxx.xxx, mhughes@xxxxxxxxx.xxx*
>
> *Subject: Diagram for WMO Statement*
>
> *Date: Tue, 16 Nov 1999 13:31:15 +0000*
>
> *Cc: k.briffa@xxxxxxxxx.xxx,t.osborn@xxxxxxxxx.xxx*
>
> *Dear Ray, Mike and Malcolm,*

Once Tim's got a diagram here we'll send that either later today or first thing tomorrow.I've just completed Mike's Nature trick of adding in the real temps to each series for the last 20 years (ie from 1981 onwards) amd from1961 for Keith's to hide the decline. Mike's series got the annua lland and marine values while the other two got April-Sept for NH land N of 20N. The latter two are real for 1999, while the estimate for 1999 for NH combined is +0.44C wrt 61-90. The Global estimate for 1999 with data through Oct is +0.35C cf. 0.57 for 1998. Thanks for the comments, Ray.

Quick briefing—the width of the tree rings used in this reconstruction vary depending on conditions—moisture, sunlight, other trees and also temperature. This particular reconstruction matched some modern records very well, and also minimized the Medieval Warming Period, making it particularly attractive to The Team, as it would counter McIntyre's criticism of Mann. Sadly, however, the tree rings showed a decline in temperature starting in 1960, just when thermometers began to gear up to show the warming we know has occurred since 1975. So 'hide the decline' does not mean anything as dramatic as a global fall in temperatures. But it's actually as important, if not more so—The Team needed to hide the decline which would raise questions about the accuracy of tree ring records. They really couldn't afford another Hockey Stick fiasco. But perhaps the saddest fact in all the Climategate scandal is that they intentionally hid this decline from politicians and policy makers, who proceeded to act in dramatic and expensive ways, confident that the tree ring record showed to perfection the long, flat history of temperatures until the recent, 'unprecedented' rise. They did this by replacing the last part of the tree ring record (which declined) with the modern instrument record (which rose).

This email and associated mails have been discussed at length since the day it came to light. For the readers of Climate Audit the issue was nothing new. Since 2005 the readers of CA had been following what was known as "the divergence problem." It had been discussed there at length. In addition, the readers were all too aware of the scientists' efforts to hide this problem in the publications of the IPCC, both in the third assessment and in the AR4 Chapter 6. But as the story broke, the principals involved, Jones and others, had a difficult time remembering what they had done. McIntyre and his band of statisticians were more than happy to lay out in detail the stunt the scientists had pulled. The readers of CA had known for some time that the scientists had hidden this problem in their graphics. And they understood why the graphics had to be altered in the IPCC reports; the mails simply provided additional evidence that the scientists knew what they were doing and they knew why they were doing it. The data that had to be hidden in the graphic was a divergence between tree rings and temperature data. That divergence was documented in the original papers. That divergence called into question the tree rings, the whole region of the world they represented and perhaps the very science that allowed them to reconstruct past temperatures.

But the graphics in the IPCC report needed to report a particular position, the position that we have some measure of certainty about past temperatures. Including a graphic with an inconvenient truth, an incontinent picture of a problem with the science would "dilute the message" So, the graphic was altered to hide this fact. "Hiding the decline" happened on more than one occasion in more than one publication, but in all cases the intention was clear. Showing the data as it appeared in the original papers would dilute the message that the scientists wanted to give to their policy making audience. So, they hid the decline. The defense they would use was disingenuous. It was simply this: The "decline," or divergence problem, was hidden in the graphics, but the references that graphic was built from discussed the problem fully.

A picture is worth a thousand words. And in their publications to policy makers they published a misleading graphic and excused this by providing references to papers that tried to explain the issue. No one but the most careful readers, and readers of CA, would catch this trick, for it would require the kind of familiarity with the underlying data that only a scientist current in the field would catch. The fact that Jones could not remember exactly what he had done is the best evidence of this. Even when Jones did finally defend what he did, he did not get it right.

In their Nov 24th Statement about the issue CRU wrote the following

> **CRU has not sought to hide the decline. Indeed, CRU has published a number of articles that both illustrate, and discuss the implications of, this recent tree-ring decline, including the article that is listed in the legend of the WMO Statement figure. It is because of this trend in these tree-ring data that we know does not represent temperature change that I only show this series up to 1960 in the WMO Statement."**

The first thing we should note about the defense is that it does not defend the actions that Jones took. The defense is a defense of CRU, not of Jones. Jones took data from a Briffa study. And following what he thought Briffa had done in his original paper he deleted data at the end of the tree ring series, data where the tree rings diverged from the temperature record, data which showed these trees no longer functioned as "treemometers." He then appended data, temperature data, to these tree rings. This is Mann's trick.

He then smoothed the data and cut it off at 1960. To excuse his behavior Jones points at the legend in the graph and he points to the publishing history of CRU. In short, Jones argues that the graphic in the WMO report is not misleading because the legend of the graphic points to an article that "explains" the divergence issue. In short, he can hide the decline because others didn't. But Jones has only defended half his action. The paper in question does not and cannot explain the trick that Jones performed to create the graphic in question. Very simply, the trick consists of deleting a portion of a tree ring series, appending real temperature data to this truncated series, and then smoothing it. It is Mann's trick. Mann's trick is not explained in the paper referred to in the legend. Briffa did not use Mann's trick. So the WMO graphic is the result of a "trick," a questionable method of handling divergent data. In his original articles Briffa did not hide this decline. In fact he displayed it fully and discussed its implications. When Jones had to create a graphic for the WMO cover, he did not present that data as the article he cited did. He performed a trick on it. A trick not documented in any literature, except the pages of CA. Showing that the graphic in question has a legend that refers to a Briffa paper does Jones no good. Briffa did not perform Mann's trick.

To get a sense of how Briffa treated the problem in the original source material, it's instructive to review his early papers, those cited by Jones, papers all too familiar to the readers of CA. In 2006 McIntyre drew attention to Briffa's discussion of the divergence problem and posted the following graphics from Briffa's paper in 1998, 2000 and 2004. First a figure from 1998 in which two measures from tree rings, ring width and maximum latewood density are shown to diverge from the recorded temperature.

Figure 15, Reconstructed Briffa 1998, Temperatures, Density and Ring Width

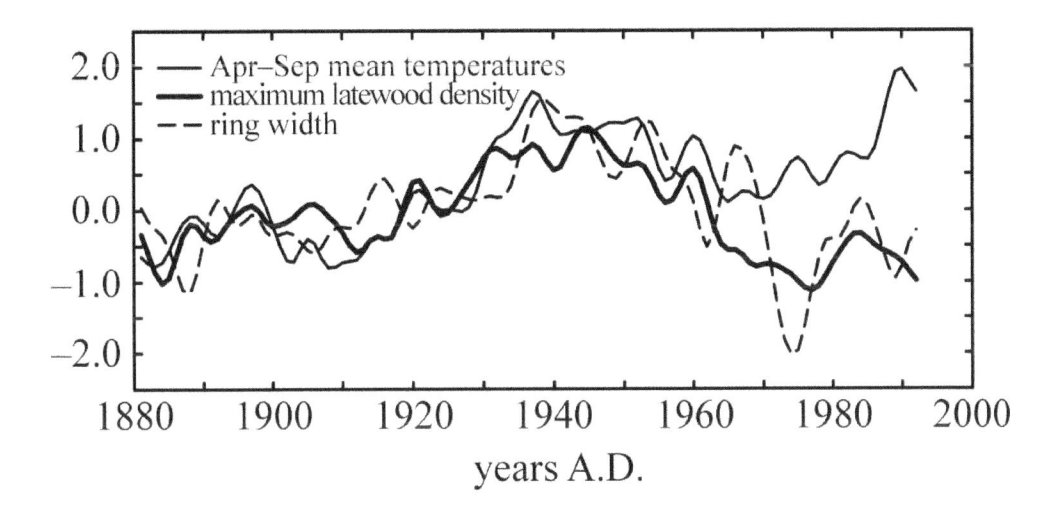

Stephen Lawrence, after Briffa 1998

And here a graphic from a paper written in 2000.

Figure 16: Reconstructed Briffa 2000, Tree Ring Density / Northern Boreal Forest

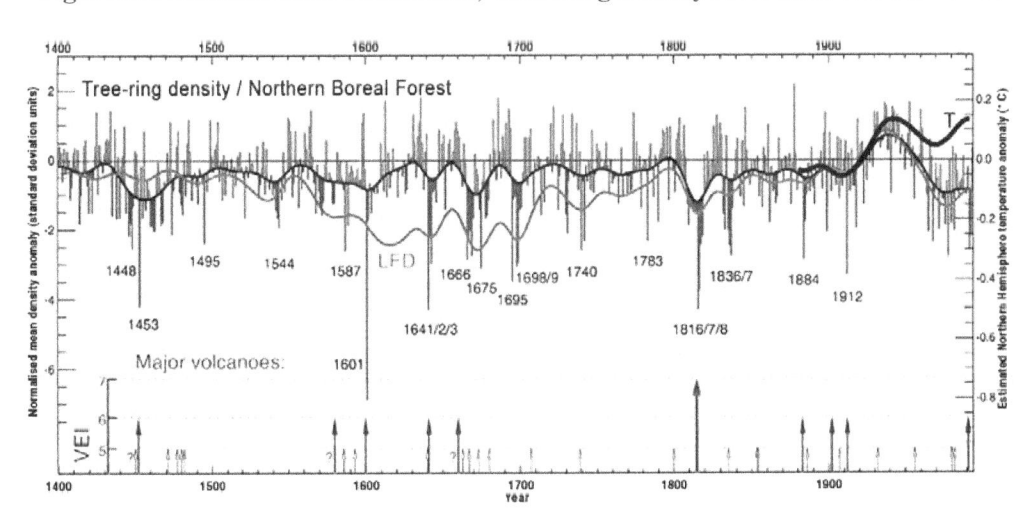

Steven Lawrence, after Briffa 2000

So, part of Jones' defense is correct. CRU, specifically, Briffa had displayed this divergence clearly in prior publications. And the underlying literature does discuss the problem in detail. Briffa, 1998 writes.

> During the second half of the twentieth century, the decadal-scale trends in wood density and summer temperatures have increasingly diverged as wood density has progressively fallen. The cause of this increasing insensitivity of wood density to temperature changes is not known, but if it is not taken into account in dendroclimatic reconstructions, past temperatures could be overestimated... In the

155

> areas where the growth data extend through to the warm late 1980s and early 1990s (NEUR, WSIB, CSIB, ESIB), the divergence is at a maximum in the most recent years. Over the hemisphere, the divergence between tree growth and mean summer temperatures began perhaps as early as the 1930s; became clearly recognisable, particularly in the north, after 1960; and has continued to increase up until the end of the common record at around 1990.

It's important to understand everything at play in Briffa's argument because it will help illustrate why Jones' graphic misleads readers. As Briffa and others had found, during the second half of the 20th century some tree measures had started to diverge from the temperature record. The underlying science holds that some trees are temperature sensitive. That is, their growth properties follow temperature, amongst other things. By studying these measures over time the scientists can reconstruct past temperatures. Very simplistically, higher temperatures correlate with thick ring width for example. Temperature goes up, tree rings get wider. (We say simplistically because the width of tree rings is influenced by a lot of factors and it is not, well, simple.) By looking at tree rings during the instrumented period and the widths seen in that period it is hoped that past temperatures can be reconstructed. The problem Briffa and others found was a divergence from this pattern. The tree ring series and the ring width series marched hand in hand from 1850 to the mid 20th century but then in some species and some locations this pattern changed.

There are several ways to handle a problem like this. If tree rings and temperatures diverge, one can question the accuracy of the temperatures, or question the tree rings, or question tree ring science itself, or some combination of these. The precedent for questioning the temperature record is established in the literature. As noted in previous chapters and on CA, Wilson and others had on occasion compiled their own temperature records, specifically in Canada, when the official record was at variance with tree rings. Briffa does not even explore this possibility with these rings. In his mind, Jones' instrument record is a fact. If the instrument record cannot be questioned then Briffa is left with fewer choices. He could just use the data as it is. But if he does this, then as he writes "past temperatures could be overestimated." That is, if the data is merely used "as is" then past temperatures will appear higher. It's vital to note that he would see such an estimation as an "over estimation."

But to some the data is just the data. If tree rings or density reflect the temperature, then these rings would indicate a warmer past than the present. To preserve a cooler past (which politically he must do to show that current warming is unprecedented), Briffa must "do something" with this data. He's left with two choices. He could argue that these tree rings show that the basic science of reconstruction is flawed. The basic science operates on a theory that a tree that is temperature sensitive today will be temperature sensitive in the past. If that's not true, if trees can sometimes function as "treemometers" and sometimes diverge from that, then the hopes of reconstructing past climate are dashed. Briffa cannot bring himself to even discuss this logical possibility. Briffa is left then with one choice: He cannot question Jones' temperature record. He cannot use the data as is and create a large MWP, he cannot question the very science he carries out, so he must "adjust" the data. In his case the adjustment is a simple deletion. The data that diverges is simply disappeared. To be sure, that deletion is "discussed" but the discussion consists merely of pointing at possible reasons for the phenomena. In 2006 McIntyre canvasses the theories in the following post:

> **Briffa et al [2002] report as follows on the problem and give what, in my opinion, is one of the most bizarre explanations – even by Hockey Team standards. In fact, I'm**

pretty sure that it was after reading this, that I wrote to Mann in April 2003 asking for his data:

Briffa et al. (1998b) discuss various causes for this decline in tree growth parameters, and Vaganov et al. (1999) suggest a role for increasing winter snowfall. We have considered the latter mechanism in the earlier section on chronology climate signals, but it appears likely to be limited to a small part of northern Siberia. In the absence of a substantiated explanation for the decline, we make the assumption that it is likely to be a response to some kind of recent anthropogenic forcing. On the basis of this assumption, the pre-twentieth century part of the reconstructions can be considered to be free from similar events and thus accurately represent past temperature variability.

As McIntyre argues this theory is no theory whatsoever. They have no explanation for the divergence, they merely assume, they have to assume, that man is somehow the cause of the decline. McIntyre then discusses Briffa 2004, where ozone is thrown up as an untested assumption. From Briffa:

The network was built over many years from trees selected to maximise their sensitivity to changing temperature.... However, in many tree-ring chronologies, we do not observe the expected rate of ring density increases that would be compatible with observed late 20th century warming. This changing climate sensitivity may be the result of other environmental factors that have, since the 1950s, increasingly acted to reduce tree-ring density below the level expected on the basis of summer temperature changes. This prevents us from claiming unprecedented hemispheric warming during recent decades on the basis of these tree-ring density data alone. Here we show very preliminary results of an investigation of the links between recent changes in MXD and ozone (the latter assumed to be associated with the incidence of UV radiation at the ground). For the time being, we circumvent this problem by restricting the calibration of the density data to the period before 1960.

The problem of the decline is "circumvented" by deleting the offending data. Since Briffa can't question the underlying science, since he can't admit a higher MWP, deleting the offending portion of the data seems his only choice. It never occurs to him the whole series of rings is suspect. Briffa:

A number of factors were taken into account when selecting the most appropriate periods for calibration and verification of the gridded density data against observed temperatures. The most important factor is the identification by Briffa et al. (1998a) of a recent downward trend in the high latitude tree-ring density data, relative to (and apparently unrelated to) warm-season temperature. This density decline becomes large enough to impair the calibration after about 1960. For this reason, both Briffa et al. (2001) and Briffa et al. (2002a) used only pre-1961 data for calibration of their subcontinental, regional temperature reconstructions. This is a reasonable choice, provided that it is explicitly stated that this approach assumes the apparent recent density decline is due to some anthropogenic factor and that similar behaviour is assumed, therefore, not to have occurred earlier in the reconstruction period – which would otherwise introduce bias in the reconstructed temperatures. At present, no satisfactory explanation of the relative MXD decline has been identified, and further work must dictate whether this assumption will be supported or rejected (Briffa et al., 1998a, 2003, and Vaganov et al., 1999, discuss and investigate possible causes).

Again in their summary, Osborn et al make it clear that they are dealing with untested assumptions:

> **The second key issue that arose during the calibration procedure is more specific to the treering density data set used here, because it relates to the decline (relative to that expected on the basis of observed summer temperatures) in density over recent decades at the high latitudes (Briffa et al., 1998). It is extremely important to try to identify the cause of this decline, though investigations are currently hampered by the early sampling of many of the sites and thus the lack of widespread data since the mid-1980s (Briffa et al., 2003). Without a satisfactory explanation, we make the untested assumption that the decline is due to an anthropogenic factor that did not occur earlier in the reconstruction period. Nevertheless, additional uncertainty must surely be associated with the reconstructions because of this assumption, particularly for earlier warm periods. The decline also complicates the reconstruction method. To prevent it from unduly influencing the calibration, the grid-box calibration was first undertaken using high-pass filtered data (the decline was removed by the filtering). The tree-ring density data were also adjusted, in an artificial way, to temporarily remove the decline; this made only a small difference in comparison with the filtered calibration, although it reduced the sensitivity of the reconstruction's mean level to the choice of calibration period and thus proved a useful, if ad hoc, way of dealing with the decline during calibration.**

As McIntyre notes, the problem of divergence is simply waved away by postulations and untested assumptions. The data diverge. If the data is included the MWP will appear warmer. So the data must be deleted and its deletion isn't justified by any tested hypothesis, it is deleted in the hopes that someone, someday will be able to explain it.

In every graphic produced by Jones and his colleagues, in the WMO graphic, in the Third IPCC report, and in the 4th IPCC, these uncertainties were brushed away, and where the underlying science was cited, **even that underlying science**, brushed the problem away. The graphics hid the decline. The referenced texts discounted the challenge to the basic science. The text did not explain either the data deletion or the marriage of the truncated tree data to instrument data. At best, the problem of divergence was explained by merely referencing the documents that raised the problem, documents that raised the problem without discussing it fully or its implications, and certainly without solving it. The graphics hide the decline and the references that supposedly explain it neither explain "Mann's trick" nor do they explain why the data diverge.

The case of hiding the decline is even deeper than this. As the mails show and as the posts at CA show, the most probable motivation for hiding the decline in the third IPCC report was to present an uncertain case as a certain one. They wanted to hide the doubt. The same issue arises in AR4 and there, as an official reviewer, McIntyre objected to it. We know this because McIntyre and his readers were able to get the AR4 reviewer comments public by sending NOAA FOIA. As a expert reviewer to AR4 McIntyre wrote:

Show the Briffa et al reconstruction through to its end; don't stop in 1960. Then comment and deal with the "divergence problem" if you need to. Don't cover up the divergence by truncating this graphic. This was done in IPCC TAR; this was misleading (comment ID #: 309-18)

McIntyre's suggestion was rejected. The graphic would remain and the solution that Briffa suggested was to discuss the divergence problem. The picture of divergence, a picture worth a thousand words, a picture which according to Jones has been addressed in many papers in many thousands of words, a picture which raises questions about the temperature record, about the science of "treemometers", about this region of the world and the trees that grow there, a picture that still has yet to be explained, is never shown.

Instead its absence is explained in Chapter 6 of AR4 in 264 words:

> **Several analyses of ring width and ring density chronologies, with otherwise well-established sensitivity to temperature, have shown that they do not emulate the general warming trend evident in instrumental temperature records over recent decades, although they do track the warming that occurred during the early part of the 20th century and they continue to maintain a good correlation with observed temperatures over the full instrumental period at the interannual time scale (Briffa et al., 2004; D'Arrigo, 2006). This 'divergence' is apparently restricted to some northern, high- latitude regions, but it is certainly not ubiquitous even there. In their large-scale reconstructions based on tree ring density data, Briffa et al. (2001) specifically excluded the post-1960 data in their calibration against instrumental records, to avoid biasing the estimation of the earlier reconstructions (hence they are not shown in Figure 6.10), implicitly assuming that the 'divergence' was a uniquely recent phenomenon, as has also been argued by Cook et al. (2004a). Others, however, argue for a breakdown in the assumed linear tree growth response to continued warming, invoking a possible threshold exceedance beyond which moisture stress now limits further growth (D'Arrigo et al., 2004). If true, this would imply a similar limit on the potential to reconstruct possible warm periods in earlier times at such sites. At this time there is no consensus on these issues (for further references see NRC, 2006) and the possibility of investigating them further is restricted by the lack of recent tree ring data at most of the sites from which tree ring data discussed in this chapter were acquired.**

Even in this abbreviated treatment of the problem it's clear that the problem raises issues with one of the principal tools of climate reconstruction. And so the issue of hiding the decline comes down to this. The graphic presented in AR4 (and various other versions) presented this as the consensus record, a busy chart that has come be known as a spaghetti chart.

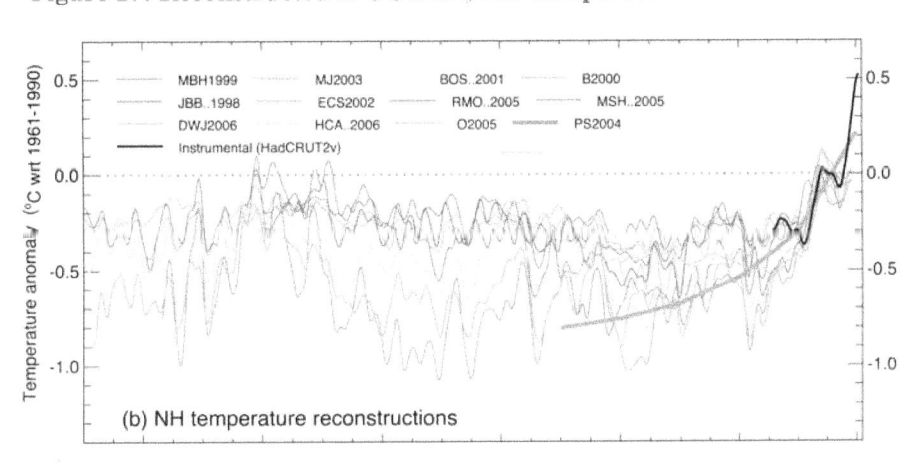

Figure 17: Reconstructed IPCC AR4, NH Temperature Reconstructions

Steven Lawrence, after IPCC AR4

And it discussed the truncation of the green lines with 264 words. Those words, however do not discuss the magnitude of the decline, they downplay its importance in the whole field, and they do not discuss potential issues with the Russian temperature data, and lastly they don't discuss how temperature data is appended to the series *prior* to smoothing. When CRU finally did post a picture after the mails were released, this is what they showed. The green line depicts CRU's November 24th version of the data:

Figure 18: Reconstructed CRU Temperature Anomaly

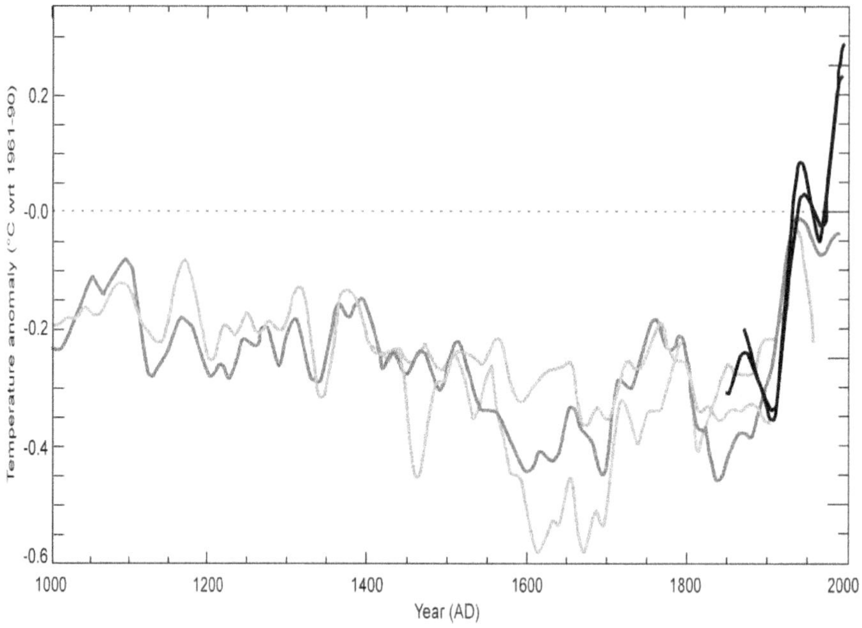

Steven Lawrence, after CRU

But even here CRU does not get the story correct. The full story can only be seen by looking at all the data Briffa deleted. The record shows that not only did the authors change the visual story, **they also deleted the underlying data from the online database at NOAA**. The community has an archive that is supposed to contain climate reconstruction data for all researchers. When the Briffa's data was archived, data past 1960 was deleted from the archive, thus insuring that anyone who used this data in the future would not have access to the inconvenient truth. It is one thing to engage in chartmanship to craft a story, but the underlying data is important. It's important for any researcher who wants to study the problem of divergence and perhaps explain what AR4 admits is unexplained today. Fortunately, the Climategate mails include all the underlying data as they were mailed to a researcher. McIntyre shows the dataset in full:

Figure 19: Reconstructed Analysis of Combined Proxy Data 1

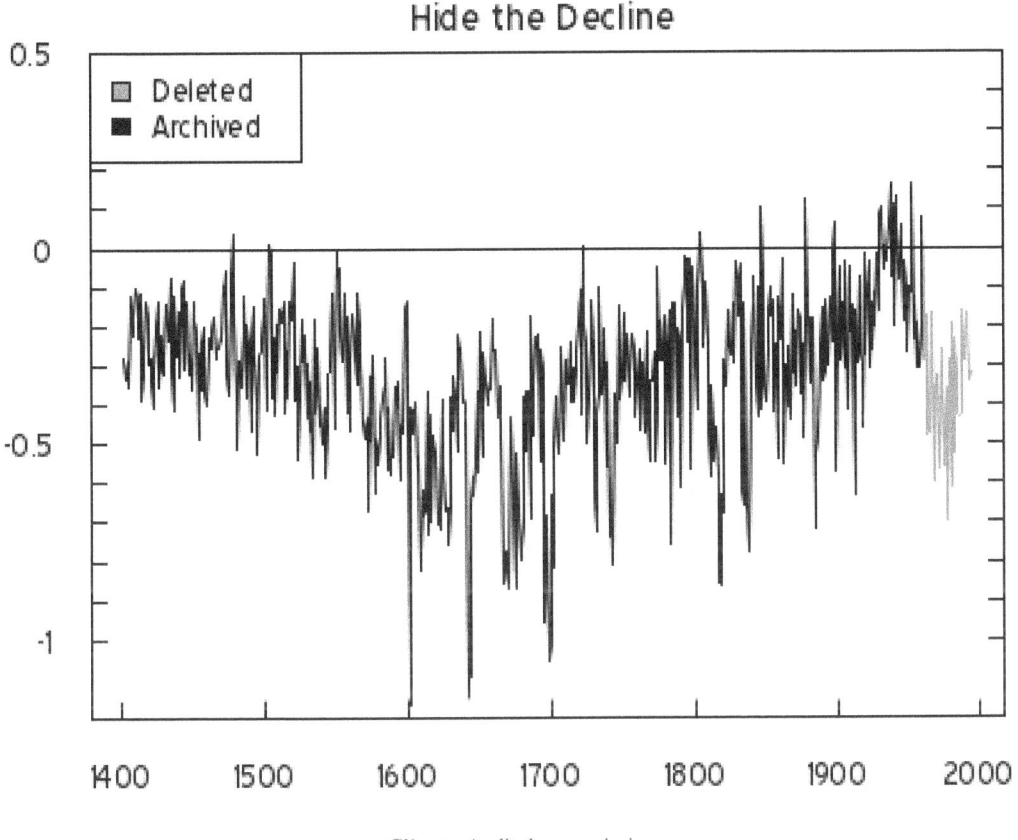

Hide the Decline

Climate Audit, by permission

So, even in November of 2009 when the mails establish that the scientists hid the decline in their graphics and knew they hid the decline, they still could not bring themselves to tell the whole story and plot all the data. Perhaps they couldn't because the source they used, the tree ring database, itself had been censored. As McIntyre points out even as of November 24th 2009, CRU is still hiding the decline. Taking the data from the archive and adding in the data from the Climategate files, McIntyre posted the picture worth a thousand words.

Figure 20: Reconstructed Analysis of Combined Proxy Data 2

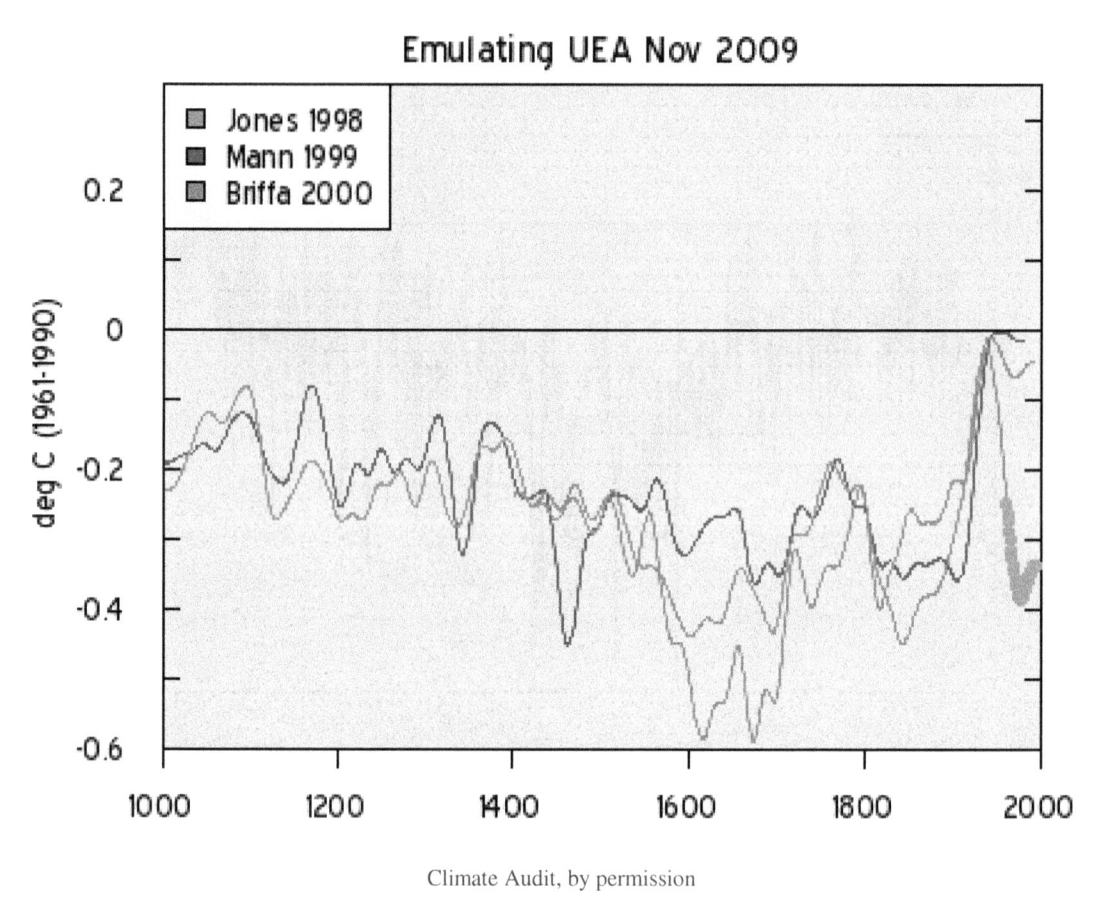

Climate Audit, by permission

The data that has been deleted from the record is shown in Magenta. The picture it paints of the divergence is different from CRU's November 24th confession, and it tells a different story than the spaghetti graph in AR4 chapter 6. And 264 words does not suffice to explain the uncertainties at play here.

The emails do not of course overturn any science. They are just emails. What they do show is a pattern of behavior. They show a group of men driven by personal, political, and institutional pressure to twist a publication system to their own ends. What they tried to conceal and spin were some fundamental uncertainties in the science. Where there was doubt they tried to downplay it because they believed "doubt" was the skeptic's product and they didn't want to give them any ammunition. As they viewed the world, skeptics were just like tobacco companies who had sold "doubt" to people about the science linking cancer and cigarettes. If skeptics were like tobacco companies, if the skeptic's product was doubt, then the last thing these scientists wanted to sell was doubt. Even if that doubt was small or of no consequence. The scientists needed to sell certainty, consensus. In the end through their actions they may have caused more uncertainty than they had tried to conceal;

Changing the questions we ask

In order to understand the subtle ways in which funding, policy interests, and personal interest twist the publication process, we need only turn to David Michaels, author of "Doubt is their product." In a Washington Post editorial he describes the corrosive effect of funding. Ulterior motives don't usually result in fraudulent studies, ulterior motives change the questions that you ask.

> **By David Michaels**
>
> **Special to The Washington Post**
>
> **Tuesday, July 15, 2008**
>
> Wal-Mart and Toys R Us announced this spring that they will stop selling plastic baby bottles, food containers and other products that contain a chemical {BPA} that can leach into foods and beverages. ... Congress is considering measures to ban the chemical. But is there enough evidence of harmful health effects on humans? One of the eyebrow-raising statistics about the BPA studies is the stark divergence in results, depending on who funded them. More than 90 percent of the 100-plus government-funded studies performed by independent scientists found health effects from low doses of BPA, while none of the fewer than two dozen chemical-industry-funded studies did. This striking difference in studies isn't unique to BPA. When a scientist is hired by a firm with a financial interest in the outcome, the likelihood that the result of that study will be favorable to that firm is dramatically increased. This close correlation between the results desired by a study's funders and those reported by the researchers is known in the scientific literature as the "funding effect."Having a financial stake in the outcome changes the way even the most respected scientists approach their research.

As Michael notes, if a scientist is hired by a firm with a financial interest, it changes the results and the structure of the science. If this holds true for companies that have a profit motive then it would seem to extend to research centers, universities, and labs who survive on the funding they get. For example the Tyndall center, discussed below, is dedicated to finding ways to adapt and mitigate the impacts of climate change. Such a mission presupposes that the climate will change in ways that are harmful. Michaels continues:

> Within the scientific community, there is little debate about the existence of the funding effect, but the mechanism through which it plays out has been a surprise.At first, it was widely assumed that the misleading results in manufacturer-sponsored studies of the efficacy and safety of pharmaceutical products came from shoddy studies done by researchers who manipulated methods and data. Such scientific malpractice does happen,but close examination of the manufacturers' studies showed that their quality was usually at least as good as, and often better than, studies that were not funded by drug companies. This discovery puzzled the editors of the medical journals, who generally have strong scientific backgrounds.

> **Richard Smith.. has written that he required "almost a quarter of a century editing . . . to wake up to what was happening." Noting that it would be far too crude, and possibly detectable, for companies to fiddle directly with results, he suggested that it was far more important to ask the "right" question. ..Smith, Bero and others have catalogued these "tricks of the trade," which include .... publishing the results of a single trial many times in different forms to make it appear that multiple studies reached the same conclusions; and publishing only those studies, or even parts of studies, that are favorable to your drug, and burying the rest.**

Here we see all the tricks employed by Mann and the rest of the Team. In drugs test, as Michael notes, companies publish the results of a single test many times to make it look like multiple studies say the same thing. In climate reconstructions McIntyre has catalogued the same trick with multiple authors using the same climate proxies and then claiming independent verification. They also, as we have seen, engage in publishing studies or data that are favorable, while burying the rest. Hiding the decline. Michael continues and discusses how "meta analysis" is created, analysis that describes the AR4 Chapter 6 process:

> **The problem is equally apparent in review articles and meta-analyses, in which an author selects a group of papers and synthesizes an overall message or pattern. Decisions about which articles to include in a meta-analysis and how heavily to weight them have an enormous impact on the conclusions. ....It has become clear to medical editors that the problem is in the funding itself. As long as sponsors of a study have a stake in the conclusions, these conclusions are inevitably suspect, no matter how distinguished the scientist.**

What Michael argues for tobacco companies and other funded science seems to clearly hold for The Team, especially in light of what the Climategate mails reveal. Michaels answer is to de link sponsorship and research and establishing research groups with an independent governing structure. For example the Health Effects Institute was established by the EPA and manufacturers. Notably the HEI

> **conducts studies paid for by corporations, but its researchers are sufficiently insulated from the sponsors that their results are credible.**

Climate science . in particular the compilation of a global temperature index, needs something similar. Soon.

Climategatekeeping

The funding effect that Michaels describes has an effect on publishing that is difficult to perceive from outside. As he notes, it took years for journal editors to figure it out. In Climategate we see the subtle effects. As Michaels notes above, the effect in meta-analysis is largely seen in the selection of articles and the weight given to them. Ammann and Wahl's 'Jesus Paper' typifies this perfectly. It goes even farther since the paper seems specifically written so that the meta analysis can use it. This takes the corruption of meta analysis to a different level. Not only are authors like

Briffa selecting their own papers and the papers of colleagues, but they are having new papers specifically written—commissioned, as it were—with the intention of having them used in the meta analysis to buttress the conclusion they want. The scientific journals then become an important battlefield. Change the science and the rules there and you can make the meta analysis say anything you want.

The approach and prejudices of the climate scientists with regards to publications can be seen most clearly in one email in a chain started by Tom Wigley at NCAR:

> At 11:53 PM 4/23/2003 -0600, Tom Wigley wrote: Dear friends, [Apologies to those I have missed who have been part of this email exchange -- although they may be glad to have been missed] I think Barrie Pittock has the right idea -- although there are some unique things about this situation. Barrie says . (1) There are lots of bad papers out there (2) The best response is probably to write a 'rebuttal' to which I add (3) A published rebuttal will help IPCC authors in the 4AR.

Wigley and others note that there are a lot of bad papers out there. The peer review process is no guarantee of truth. But Wigley adds a crucial motive: helping the authors of the AR4, which as we saw in the case of Chapter 6 involved shoehorning papers into the process. Ordinarily, bad science will merely wither and die. But here Wigley and others have an agenda driven by a process.

Wigley then lays out his hierarchy of values: laudably he sees correcting bad science as the first concern, second is personal motives. They take their science first, but interestingly they take the personal attacks as next in importance. Reputation matters to these men. And third, they consider the political agenda of their opponents.

> Let me give you an example. There was a paper a few years ago by Legates and Davis in GRL (vol. 24, pp. 2319-1222, 1997) that was nothing more than a direct and pointed criticism of some work by Santer and me -- yet neither of us was asked to review the paper. We complained, and GRL admitted it was poor judgment on the part of the editor. Eventually (2 years later) we wrote a response (GRL 27, 2973-2976, 2000). However, our response was more than just a rebuttal, it was an attempt to clarify some issues on detection. In doing things this way we tried to make it clear that the original Legates/Davis paper was an example of bad science (more bluntly, either sophomoric ignorance or deliberate misrepresentation). Any rebuttal must point out very clearly the flaws in the original paper. If some new science (or explanations) can be added -- as we did in the above example -- then this is an advantage.
> There is some personal judgment involved in deciding whether to rebut. Correcting bad science is the first concern. Responding to unfair personal criticisms is next. Third is the possible misrepresentation of the results by persons with ideological or political agendas. On the basis of these I think the Baliunas paper should be rebutted by persons with appropriate expertise. Names like Mann, Crowley, Briffa, Bradley, Jones, Hughes come to mind. Are these people willing to spend time on this?

Wigley continues with another example. And we watch how sensitive he is to the review process, the selection of reviewers, and the assumptions that inform his conclusion:

> There are two other examples that I know of where I will probably be involved in writing a response. The first is a paper by Douglass and Clader in GRL (vol. 29, no. 16,

> *10.1029/2002GL015345, 2002). I refereed a virtually identical paper for J. Climate, recommending rejection. All the other referees recommended rejection too. The paper is truly appalling -- but somehow it must have been poorly reviewed by GRL and slipped through the net. I have no reason to believe that this was anything more than chance. Nevertheless, my judgment is that the science is so bad that a response is necessary. The second is the paper by Michaels et al. that was in Climate Research (vol. 23, pp. 19, 2002). Danny Harvey and I refereed this and said it should be rejected. We questioned the editor (deFreitas again!) and he responded saying The MS was reviewed initially by five referees. The other three referees, all reputable atmospheric scientists, agreed it should be published subject to minor revision. Even then I used a sixth person to help me decide. I took his advice and that of the three other referees and sent the MS back for revision. It was later accepted for publication. The refereeing process was more rigorous than usual. On the surface this looks to be above board -- although, as referees who advised rejection it is clear that Danny and I should have been kept in the loop and seen how our criticisms were responded to. It is possible that Danny and I might write a response to this paper -- deFreitas has offered us this possibility. This second case gets to the crux of the matter. I suspect that deFreitas deliberately chose other referees who are members of the skeptics camp. I also suspect that he has done this on other occasions.*

When we consider how Chapter 6 of the IPCC's AR4 was written, when we consider how the criticisms of McIntyre were answered by appealing to unpublished literature, when we consider that Briffa and others were reviewing their own work, how the journal system was manipulated, then we can be as skeptical of Chapter 6 as Wrigley is of deFreitas. He continues:

> *How to deal with this is unclear, since there are a number of individuals with bona fide scientific credentials who could be used by an unscrupulous editor to ensure that 'anti-greenhouse' science can get through the peer review process (Legates, Balling, Lindzen, Baliunas, Soon, and so on). The peer review process is being abused, but proving this would be difficult. The best response is, I strongly believe, to rebut the bad science that does get through. Jim Salinger raises the more personal issue of deFreitas. He is clearly giving good science a bad name, but I do not think a barrage of ad hominem attacks or letters is the best way to counter this. If Jim wishes to write a letter with multiple authors, I may be willing to sign it, but I would not write such a letter myself. In this case, deFreitas is such a poor scientist that he may simply disappear. I saw some work from his PhD, and it was awful (Pat Michaels' PhD is at the same level).*
> *_____ Best wishes to all, Tom.*

Here Wrigley is discussing writing a letter to criticize deFreitas, to delegitimize him. As he points out however, deFreitas, like bad science may just wither away. The point of course is that Wigley and others feel that the usual course of things may not be sufficient. Banishment may be in order. Ironically, while Wigley considers this, he exposes his personal interest.

Mann responds, agreeing with Wigley and offering a political angle as well: It's interesting to note that he considers what skeptics say behind closed doors as relevant. He considers it relevant that they make ad hominem attacks on IPCC scientists. By that standard, by Mann's standard, he stands self condemned by the Climategate files.

Michael E. Mann wrote:

Dear Tom et al, Thanks for comments--I see we've built up an impressive distribution list here! This seemed like an appropriate point for me to chime in here. By in large, I agree w/ Tom's comments ......This was the basis for their press release arguing for a "MWP" that was "warmer than the 20th century" (a non-sequitur even from their awful paper!) and for their bashing of IPCC and scientists who contributed to IPCC (which, I understand, has been particularly viscious and ad hominem inside closed rooms in Washington DC where their words don't make it into the public record). This might all seem laughable, it weren't the case that they've gotten the (Bush) White House Office of Science & Technology taking it as a serious matter (fortunately, Dave Halpern is in charge of this project, and he is likely to handle this appropriately, but without some external pressure).

Mann, like Wigley, thinks the counter attack has to go beyond merely correcting or ignoring the bad science. He believes that the skeptics have compromised the peer review process.

So while our careful efforts to debunk the myths perpetuated by these folks may be useful in the FAR, they will be of limited use in fighting the disinformation campaign that is already underway in Washington DC. Here, I tend to concur at least in sprit w/ Jim Salinger, that other approaches may be necessary. I would emphasize that there are indeed, as Tom notes, some unique aspects of this latest assault by the skeptics which are cause for special concern. This latest assault uses a compromised peer-review process as a vehicle for launching a scientific disinformation campaign (often viscious and ad hominem) under the guise of apparently legitimately reviewed science, allowing them to make use of the "Harvard" moniker in the process.

And in the press coverage Mann also is quick to focus in on the political slant and like a political activist Mann and people who think like him plan an organized "protest" or mass resignation.

Fortunately, the mainstream media never touched the story (mostly it has appeared in papers owned by Murdoch and his crowd, and dubious fringe on-line outlets). Much like a server which has been compromised as a launching point for computer viruses, I fear that "Climate Research" has become a hopelessly compromised vehicle in the skeptics' (can we find a better word?) disinformation campaign, and some of the discussion that I've seen (e.g. a potential threat of mass resignation among the legitimate members of the CR editorial board) seems, in my opinion, to have some potential merit. Like a This should be justified not on the basis of the publication of science we may not like of course, but based on the evidence (e.g. as provided by Tom and Danny Harvey and I'm sure there is much more) that a legitimate peer-review process has not been followed by at least one particular editor.

And here Mann's suggestion can be turned on the climate scientists themselves, as the Climategate files show that they sought to bend the rules of the peer review process and the process of writing Chapter 6 of the AR4. Mann concludes this mail with the suggestion of getting someone to their liking in editorial positions.

Incidentally, the problems alluded to at GRL are of a different nature--there are simply too many papers, and too few editors w/appropriate disciplinary expertise, to get many of the papers submitted there properly reviewed. It's simply hit or miss with respect to whom the chosen editor is. While it was easy to make sure that the worst papers, perhaps including certain ones Tom refers to, didn't see the light of the day at /J. Climate/, it was inevitable that such papers might slip through the cracks at e.g. GRL--there is probably little that can be done here, other than making sure that some qualified and responsible climate scientists step up to the plate and take on editorial positions at GRL.

best regards,   Mike

Mark Eakin of NOAA suggests a letter to OSTP, the Office of Science and Technology Policy. In short the scientists saw themselves in a battle against political forces and they involved themselves in the process of making policy. That opens up the question of their political motives in the same way they wanted to open up the question of their opponents' political motives.

At 09:27 AM 4/24/03 -0600, Mark Eakin wrote:

At this point the question is what to do about the Soon and Baliunas paper. Would Bradley, Mann, Hughes et al. be willing to develop an appropriate rebuttal? If so, the question at hand is where it would be best to direct such a response. Some options are: 1) A rebuttal in Climate Research 2) A rebuttal article in a journal of higher reputation 3) A letter to OSTP   The first is a good approach, as it keeps the argument to the level of the current publication. The second would be appropriate if the Soon and Baliunas paper were gaining attention at a more general level, but it is not. Therefore, a rebuttal someplace like Science or Nature would probably do the opposite of what is desired here by raising the attention to the paper. The best way to take care of getting better science out in a widely read journal is the piece that Bradley et al. are preparing for Nature. This leaves the idea of a rebuttal in Climate Research as the best published approach. A letter to OSTP is probably in order here. Since the White House has shown interest in this paper, OSTP really does need to receive a measured, critical discussion of flaws in Soon and Baliunas' methods. I agree with Tom that a noted group from the detection and attribution effort such as Mann, Crowley, Briffa, Bradley, Jones and Hughes should spearhead such a letter. Many others of us could sign on in support. This would provide Dave Halpern with the ammunition he needs to provide the White House with the needed documentation that hopefully will dismiss this paper for the slipshod work that it is. Such a letter could be developed in parallel with a rebuttal article. I have not received all of the earlier e-mails, so my apologies if I am rehashing parts of the discussion that might have taken place elsewhere.   Cheers, Mark

Mann replies to a mailing list that includes the chairman of the IPCC, editors of journals, and fellow scientists, a real Who's Who of the climate change world, including James Hansen, Steven Schneider, and the head of the Intergovernmental Panel on Climate Change, Rajenda Pachauri:

From: "Michael E. Mann" <mann@multiproxy.evsc.virginia.edu>   To: mark.eakin@noaa.gov   Subject: Re: My turn   Date: Thu, 24 Apr 2003 12:39:14 -0400   Cc: Tom Wigley <wigley@ucar.edu>, Phil Jones <p.jones@uea.ac.uk>, Mike Hulme <m.hulme@xxx.xx.xx>, Keith Briffa <k.briffa@xxx.xx.xx>, James Hansen <jhansen@giss.nasa.gov>, Danny Harvey <harvey@cirque.geog.utoronto.ca>, Ben Santer <santer1@llnl.gov>, Kevin Trenberth <trenbert@ucar.edu>, Robert wilby <rob.wilby@kcl.ac.uk>, Tom Karl <Thomas.R.Karl@noaa.gov>, Steve Schneider <shs@stanford.edu>, Tom Crowley <tcrowley@duke.edu>, jto <jto@u.arizona.edu>, "simon.shackley" <simon.shackley@umist.ac.uk>, "tim.carter" <tim.carter@vyh.fi>, "p.martens" <p.martens@icis.unimaas.nl>, "peter.whetton" <peter.whetton@dar.csiro.au>, "c.goodess" <c.goodess@xxx.xx.xx>, "a.minns" <a.minns@xxx.xx.xx>, Wolfgang Cramer <Wolfgang.Cramer@pik-potsdam.de>, "j.salinger" <j.salinger@niwa.co.nz>, "simon.torok" <simon.torok@csiro.au>, Scott Rutherford <srutherford@deschutes.gso.uri.edu>, Neville Nicholls <n.nicholls@bom.gov.au>, Ray Bradley <rbradley@geo.umass.edu>, Mike MacCracken <mmaccrac@comcast.net>, Barrie Pittock <Barrie.Pittock@csiro.au>, Ellen Mosley-Thompson <thompson.4@osu.edu>, "pachauri@teri.res.in" <pachauri@teri.res.in>, "Greg.Ayers" <Greg.Ayers@csiro.au>, wuebbles@atmos.uiuc.edu, christopher.d.miller@xxxx.xxx, mann@virginia.edu <x-flowed>

HI Mark, Thanks for your comments, and sorry to any of you who don't wish to receive these correspondances.. Indeed, I have provided David Halpern with a written set of comments on the offending paper(s) for internal use, so that he was armed w/ specifics as he confronts the issue within OSTP. He may have gotten additional comments from other individuals as well--I'm not sure. I believe that the matter is in good hands with Dave, but we have to wait and see what happens. In any case, I'd be happy to provide my comments to anyone who is interested. I think that a response to "Climate Research" is not a good idea. Phil and I discussed this, and agreed that it would be largely unread, and would tend to legitimize a paper which many of us don't view as having passed peer review in a legitimate manner. On the other hand, the in prep. review articles by Jones and Mann (Rev. Geophys.), and Bradley/Hughes/Diaz (Science) should go a long way towards clarification of the issues (and, at least tangentially, refutation of the worst of the claims of Baliunas and co). Both should be good resources for the FAR as well... cheers, mike

The pattern is clear through various mails. In order to combat what they took to be bad science promoted in journals with questionable credentials, bad science which they thought was motivated by skeptics' personal defects, motivated by skeptics' financial and political interests, the climate scientists went beyond merely combating bad science with good science. They were not blind to their own personal interests, their own financial interests, and their own political interests. Rather than combating bad science with good science they fought against transparency in science and worked to corrupt the scientific publishing system. They became the enemy they feared. Toward the end, Mann again gives us insight into the paranoia:

From: Michael Mann <mann@xxxxxxxxx.xxx> To: Phil Jones <p.jones@xxxxxxxx.xxx> Subject: Re: attacks against Keith Date: Wed, 30 Sep 2009 11:06:20 -0400 Cc: Gavin Schmidt <gschmidt@xxxxxxxxx.xxx>, Tim Osborn <t.osborn@xxxxxxxxx.xxx>

Hi Phil, lets not get into the topic of hate mail. I promise you I could fill your inbox w/ a very long list of vitriolic attacks, diatribes, and threats I've received. Its part of the attack of the corporate-funded attack machine, i.e. its a direct and highly Intended outcome of a highly orchestrated, heavily-funded corporate attack campaign. We saw it over the summer w/ the health insurance industry trying to defeat Obama's health plan, we'll see it now as the U.S. Senate moves on to focus on the cap & trade bill that passed congress this summer. It isn't coincidental that the original McIntyre and McKitrick

E&E paper w/ press release came out the day before the U.S. senate was considering the McCain Lieberman climate bill in '05. we're doing the best we can to expose this. I hope our Realclimate post goes some ways to exposing the campaign and pre-emptively deal w/ the continued onslaught we can expect over the next month. thanks for alerting us to that detail of Kaufman et al which I'd overlooked.

Climate Audit, unlike Mann's twisted depiction, isn't funded by corporations. It's a self funded volunteer project. There is no highly orchestrated attack machine. When the Climategate mails came into the possession of WUWT and Mosher there was no call out to a public relations firm to craft a message or plan a media campaign. Part of Real Climate's ineffectiveness is this continued belief. They are fighting an enemy that doesn't exist. The right-wing think tanks that do publish on climate change are slow, cumbersome and follow the agenda set by independent skeptic sites—they do not set the agenda.

The pressure the climate scientists put on the journal process extends beyond threatening editors and influencing reviewers; in some cases they worked to keep critical articles out of the journals.

As we have shown there were two questions or doubts about climate science that concerned the scientists most. Doubts about reconstructions and doubts about UHI. Not unsurprisingly these were two main thrusts for FOIA. And not surprising the scientists' questionable behavior in publishing center around these two issues. We turn again to the question of UHI.

The major doubts about the surface station record are not that great. It's largely agreed that the record shows warming. In his work Jones and CRU had estimated this at less than 10% of all warming seen in the record. Even the most ardent skeptics argue would only attribute roughly 50% of the warming over land to UHI. This view of UHI is best expressed by Ross McKitrick and Pat Michael's 2004 paper which argues that UHI may be at the root of increased temperatures over land. In a later paper McKitrick would estimate the effect of UHI at no more than 50% of the figure seen over land. (The ocean doesn't suffer from UHI.)

> **Monthly surface temperature records from 1979 to 2000 were obtained from 218 individual stations in 93 countries and a linear trend coefficient determined for each site. This vector of trends was regressed on measures of local climate, as well as indicators of local economic activity (income, GDP growth rates, coal use) and data quality. The spatial pattern of trends is shown to be significantly correlated with non-climatic factors, including economic activity and sociopolitical characteristics of the region. The analysis is then repeated on the corresponding IPCC gridded data, and very similar correlations appear, despite previous attempts to remove non-climatic effects. The socioeconomic effects in the data are shown to add up to a net warming bias, although more precise estimation of its magnitude will require further research.**

Phil Jones' reply to this challenge to his position on UHI:

From: Phil Jones <p.jones@xxxxxxxxx.xxx>

To: "Michael E. Mann" <mann@xxxxxxxxx.xxx>

Subject: HIGHLY CONFIDENTIAL

Date: Thu Jul 8 16:30:16 2004

Mike,

…..The other paper by MM [Mckittrick and Michels] is just garbage - as you knew. De Freitas again. Pielke is also losing all credibility as well by replying to the mad Finn as well - frequently as I see it. I can't see either of these papers being in the next IPCC report. Kevin and I will keep them out somehow - even if we have to redefine what the peer-review literature is !

Cheers

Phil

Jones actually carried through on his threat as McIntyre made clear in a post made after this mail became public. Jones refused to include Mckitrick's 2004 paper in either the first draft or second draft of the AR4, relenting only when McKitrick pointed out that there was another paper showing the same results. Even then, Jones refused to treat the article fairly and dismissed it by making up arguments out of whole cloth.

> **The paper in question (McKitrick and Michaels Clim Res 2004) … was submitted in July 2003 and accepted on Apr 20, 2004. McKitrick and Michaels submitted what Jones later calls an "expanded" version of this paper to International Journal of Climatology in May 2004, which was then assigned to Andrew Comrie of the University of Arizona. Comrie sought a review from the omnipresent Phil Jones (and apparently two others). The submission was rejected. …**
>
> **Contrary to the spin …, it is a matter of fact that Trenberth and Jones kept Michaels and McKitrick (2004) out of the AR4 First Draft. (I searched and confirmed this.) As an IPCC peer reviewer, McKitrick and another reviewer (Vincent Grey) vigorously objected to the exclusion.**
>
> **Trenberth and Jones flatly rejected their comments. The following is one example. Consult the AR4 First Order Draft Review Comments for others.**
>
> *References are plentiful. Those of value are cited Rejected. The locations of socioeconomic development happen to have coincided with maximum warming, not for the reason given by McKitrick and Michaels (2004) but because of the strengthening of the Arctic Oscillation and the greater sensitivity of land than ocean to greenhouse forcing owing to the smaller thermal capacity of land.*
>
> **Ross tells me that there was no peer reviewed literature at the time (or to this day) specifically supporting the Trenberth and Jones attribution of the effect to the "strengthening of the Arctic Oscillation".**
>
> **In the Second Order Draft, Trenberth and Jones were once again successful in keeping Michaels and McKitrick (2004) out of the IPCC Draft. Once again, as IPCC peer reviewers, McKitrick and Grey objected and once again, the Trenberth and Jones Author Responses were dismissive. For example:**
>
> *Rejected. McKitrick and Michaels (2004) is full of errors. There are many more papers in support of the statement than against it.*

> Or again:
>
> The locations of socioeconomic development happen to have coincided with maximum warming, not for the reason given by McKitrick and Michaels (2004) but because of the strengthening of the Arctic Oscillation and the greater sensitivity of land than ocean to greenhouse forcing owing to the smaller thermal capacity of land.
>
> Readers who wish to canvass all the comments can search the Review Comments at the above links.
>
> However, there was a complication for Jones and Trenberth, who had thus far been successful in carrying out their threat. This time, there was a second article (de Laat and Maurelis. IJC 2006) making very similar arguments to McKitrick and Michaels.....
>
> This time, Trenberth and Jones grudgingly agreed to mention the two articles in the IPCC report. However, they accompanied the mention with an extremely dismissive characterization – a characterization which (1) was made without any citation to peer reviewed literature and (2) that had not itself been submitted to external IPCC peer reviewers; and (3) to which McKitrick and Michaels had no previous opportunity to reply.

The argument over UHI is over a few tenths of a degree. Releasing the data and code should have resolved this. Since 1980 Jones estimates the warming over land to be about .3C per decade or .6C in total. Mckitrick estimates it at .34C in total. And the land is only 30% of the globe. But Jones and CRU fought this for years and have only recently relented. With allegations from the Russians on the table that CRU have misrepresented the record in Russia it is questionable if any analysis from CRU will be believed without a complete audit. Finally, the emails detail a paper that Jones had spiked at a journal. That paper addressed the UHI problem in Russia had results that Jones did not like.

What we see by examining the whole record in context is: A process of journal publishing that was corrupted to serve a IPCC process that was also corrupted, and the result was a record of the science that asserts more certainty than the underlying science supports. The scientists and those who support them obviously see their cause as a noble one. They see a planet in danger. But to make their case seem stronger than it is they "worked" the publishing machine. What they overlooked of course was the counter publishing machine: the internet, which has tried to break down subscription walls, speed up the process and open up participation. Climategate has highlighted each of these factors and has led to calls for a complete revamping of how the process of publication works.

We wrote earlier about the culture shock that was evident when The Team first encountered the quick pace and marked lack of respect for institutional authority that is prevalent in the blogosphere. Indeed, many of their problems seem to stem from their insistence on a 'command and control' approach to communications, which has consistently wrong-footed them when dealing with the more chaotic world of the internet.

Possible Motives

The motives that drove the scientists to this behavior are probably varied and are still unknown, but some pressures that influence these motives are clear.

In academic life, funding is a perpetual worry for senior academics and administrators. Institutions of higher education are very concerned with receiving outside funding for departments and projects, and academics are under continuous pressure to apply successfully for grants that will fund, not just their research, but elements of the academic infrastructure that supports that research. Grant-writing is one of the supreme academic skills.

Getting government funding for a longitudinal research project is the Holy Grail, and one doesn't have to think a scientist is a crook to understand that writing proposals directed to areas of government interest is a natural consequence of these pressures. And government interest in climate change has been growing for twenty years. So, for example, Phil Jones' ability to win significant funding from the United States Department of Energy for continuing research programs at the UK's East Anglia University's Climate Research Unit is a windfall that is hugely beneficial to the University as a whole.

The role that money and finances play in Climategate is not the role that money typically plays in corruption cases. There is no question of individuals presenting scientific views merely for personal gain. Nothing in the files remotely suggests science for hire. The influence of money is more subtle than that. To be sure, there are ironic mails that indicate that CRU and other organizations benefitted from the largess of corporations, in fact the same corporations that Mann had imagined were funding Climate Audit were actually contributing to CRU: major oil companies:

> *From: "Mick Kelly" <m.kelly@xxxxxxxxx.xxx>*
>
> *To: m.hulme@xxxxxxxxx.xxx*
>
> *Subject: Shell*
>
> *Date: Wed, 05 Jul 2000 13:31:00 +0100*
>
> *Reply-to: m.kelly@xxxxxxxxx.xxx*
>
> *Cc: t.oriordan@xxxxxxxxx.xxx, t.o'riordan@xxxxxxxxx.xxx*
>
> *Mike*
>
> *Had a very good meeting with Shell yesterday. Only a minor part of the agenda, but I expect they will accept an invitation to act as a strategic partner and will contribute to a studentship fund though under certain conditions. I now have to wait for the top-level soundings at their end after the meeting to result in a response. We, however, have to discuss asap what a strategic partnership means, what a studentship fund is, etc, etc. By*

email? In person? I hear that Shell's name came up at the TC meeting. I'm ccing this to Tim who I think was involved in that discussion so all concerned know not to make an independent approach at this stage without consulting me! I'm talking to Shell International's climate change team but this approach will do equally for the new foundation as it's only one step or so off Shell's equivalent of a board level. I do know a little about the Fdn and what kind of projects they are looking for. It could be relevant for the new building, incidentally, though opinions are mixed as to whether it's within the remit. Regards

Mick

There is nothing in the mails to suggest that the climate science created with funding from corporate interests is necessarily wrong. The science should stand or fall on its own regardless of the source of funding. The same of course should hold true for skeptical views funded by other sources. The point is this. If there is transparency in the science, if the data and methods are made available to be checked, then the motives of the funding source becomes a non-issue. Drug companies make money from their science. The fact that they make a profit doesn't make their medicines ineffective or unsafe. The check against the profit motive is, of course independent checking of the results. Specifically those results are check by people with different motives. Climate science needs such a review.

The funding does however appear to play a role in the selection of which science is done. Funding changes the questions you ask. That is, the science is focused on proving the case of global warming rather than challenging it and so there is the ever present danger of confirmation bias. At one point in the mails Mike Hulme discusses a proposal that UEA will eventually succeed in:

Mike Hulme <m.hulme@xxx.xx.xx> 09/28/99 02:34AM >>>

Dear Sujata,

This may well not be news to you, but the UK government has recently requested bids from UK universities to house a new 'National Climate Change Centre'. The Centre would receive funds of 2 million pounds sterling per year for (at least initially) five years. The role of the Centre would be to compliment existing work on climate modelling and data analysis (IPCCWGI areas) by focussing on 'solutions' (mitigation and adaptation options and their implementation), specifically for the UK government and business community, but within a global context. The emphasis appears to be on IPCCWG3 area with a strong commitment to integrated research, but with some overlap with WG2. The Centre would carry out independent research, but would also be expected to make use of, and to integrate, exisiting UK research and expertise. It would be expected to contribute to and to foster interdisciplinary research that underpins sustainable solutions to the climate change problem. UEA is making a bid for this Centre. Applications are due by mid-October. UEA is well-known for CRU, but it also has strengths in data distribution to the climate impacts community, in impacts research, and in environmental economics (CSERGE). While these areas are fundamental foundation stones for the science that the Centre is expected to develop, the Centre would need to expand significantly beyond these areas. We have a Consortium in place as follows- 6-7 Senior Partners - (UEA, UMIST, U.Southamton, Dept. Economics at U.Cambridge, Cranfield, Leeds Institute of

Transport Studies, IH and ITE) - Affiliated UK Organisations - (we have 6-8 of these)- Supporting Business Links- Supporting International Organisations If UEA were to succeed in its bid for the Centre, then it would seek to develop strong links with other institutions abroad in order to strengthen its own intellectual base and, through such links, to contribute to the development and implementation of the science. We would see TERI as one of these Supporting International Organisations.To this end, we would like a short letter of support from yourself – on behalf of the Policy Analysis Division, or a wider TERI grouping if you feel able to represent them - indicating that you fully support the UEA bid and would exclusively lend your backing to this Consortium and be keen to interact closely with us at a research level were the Centre to come to UEA. This interaction may take the form of exchanging scientists, testing out new methodologies, developing/advising on workshops, providing entry-points into international policy initiatives, etc., etc.Nothing too formal or lengthy at this stage, but we would like to provide the Council's with a flavour of the breadth of our existing and future colloboration in the field and our ability to mobilise support in our favour.Many thanks. Please send to Prof. Trevor Davies, Dean, Environmental Sciences, UEA, Norwich, NR4 7TJ, before the 12th October.Feel free to ask me for more details, etc. Our written text is beginning to take shape and we will circulate a draft of this to you before the bid goesin.Regards,Mike

The aim of this center presupposes that climate change is a reality. It's not an open question. And so the science it funds at UEA will necessarily not be able to question certain assumptions. To the extent that UEA funding depends upon the truth of global warming, scientists within UAE cannot question it without undermining their financial well being.

As to other motives—fame, ambition, ego, etc., we choose not to speculate. People differ, and are different at different times. The likeliest guess is that most of them thought they were doing the best they could at any given moment, but their best simply wasn't good enough to meet the demands of a global policy issue.

Policy Impacts

The policy implications of Climategate exist, but amount to nothing when compared with the long-term impacts of the mindset and behavior that Climategate revealed. Climategate by itself is a scandal, a nine-day wonder that will be replaced in the headlines quickly enough that we have our eyes on the clock—not the calendar—as we write this. It may have had a minor impact on COP-15, the climate summit recently concluded in Copenhagen, but only because it was combined with frequent reporting on the number of private planes and limos used by delegates. The policy impacts of the decades-long behavior that is exposed in the Climategate emails is profound.

The most important impact is quite likely to be a lessening of trust in science as a means for exploring the truth about our world. Scientists can and do sometimes take off their lab coat and put on a policy advocate's hat—and as long as they're clear about it, there is no cause for complaint. But the idea that they can become back-room agitators for a political position and shape their science to help their cause is devastating, and the world will probably pay a heavy price for this in the years to come.

Another impact is likely to be a loss of focus and political capital for the Obama administration. The focus on disaster scenarios to scare us into drastic action diverted us from taking the next right actions. As with the funding effect, it's all about the questions you ask.

President Barack Obama assumed office with only two major crises and one imagined crisis--the economy, which is a house afire, and healthcare, which is a sinking ship, and global warming which is a boy crying wolf. As we hope we have shown, the fanatical determination of some scientists, lobbyists and politicians to keep a catastrophic scenario of global warming before the public eye and on the political agenda involved a lot of chicanery. They colored the presentation of uncertainties in IPCC reports. They tried to game the peer review process, perhaps even to the extent of having an unsympathetic journal editor replaced. They deleted emails, told others to delete emails and threatened to delete data files--data files which have since disappeared.

And they trashed other scientists, trampling them underfoot if they showed any sign of independence or disagreement with their party line--anything to do with the environment had to be CO2, all CO2, twenty four hours a day. Consequently, the administration was forced into expending political time and capital into solving what looks to be like a faux crisis.

But there are broader implications. If President Obama had not felt constrained to make an omnibus energy package his second priority (after the economy), it is certainly possible that healthcare would be a done deal by now. Obviously that's speculation, but the tireless drumbeat of alarmist propaganda, inspired by the scientists whose emails we're all reading, means we will never know. Their hysterical hockey sticks, as translated by the non-scientists who tried to panic the public, succeeded in keeping global warming at the top of the political agenda, even while it was at the bottom of public concerns. President Obama lost out because of this--and perhaps so did we all.

Or take biofuels. Global commodity prices, and in particular the price of food, have pushed tens of millions back into poverty, because in our race to find a fuel that was friendlier to the environment, we began to convert corn into fuel. Had we adopted a saner approach to dealing with the effects of global warming, we wouldn't be standing by while Archer Daniels Midland gets richer while the poorest of the poor get poorer.

And would an energy policy not fueled by hysteria really be pushing so hard for windmills? Wouldn't we be looking at smaller, more efficient ways to move towards green energy generation? And what of nuclear? If we look at the opponents of nuclear, isn't their product doubt?

As for research--if you think of the $50 to $70 billion reputedly spent on climate research over the past decade, can you imagine if we had invested that into research on clean energy instead?

A well-financed campaign, slickly produced and artfully marketed, has attempted to influence world policy on environmental issues. Global warming activists like to make the claim that that campaign was run by skeptics financed by big oil against measures to curb global warming. But that does not appear to be the case. The big money campaign, financed by NGOs and big energy companies like General Electric, spent 100 times as much money trying to convince us all that the scientists who were hiding and changing data, playing pathetic political games and hobnobbing with the great and the good on their junkets to climate conferences around the world, were rock solid with the science and dead on with their predictions of disaster.

CHAPTER NINE: AFTER THE GOLD RUSH

As we have written, home base on the Internet for those who believe in global warming is a weblog called Real Climate. It is moderated by NASA scientist Gavin Schmidt, and most of The Team are listed as contributors to the weblog. Real Climate has served as the unofficial voice of global warming science for years, and it has been characterized by a ruthless willingness to shut off discussion by deleting comments or not allowing them to appear. This tendency got so out of hand that two websites appeared with the sole purpose of hosting comments that had been deleted from Real Climate.

Almost everyone in the tight little world of Internet climate junkies (including both your humble authors) was waiting for Real Climate to respond to a scandal that was spreading past the blog frontier and into major media waters. And on November 20, just three days after the release of the emails and files, Gavin Schmidt posted a long screed called 'The CRU Hack.' Keeping a very British stiff upper lip, Schmidt planted several stakes in the ground, establishing reference points that global warming bloggers and sympathetic souls in major media outlets would be using for days to come.

"More interesting is what is not contained in the emails. There is no evidence of any worldwide conspiracy, no mention of George Soros nefariously funding climate research, no grand plan to 'get rid of the MWP', no admission that global warming is a hoax, no evidence of the falsifying of data, and no 'marching orders' from our socialist/communist/vegetarian overlords."

..." it's important to remember that science doesn't work because people are polite at all times. Gravity isn't a useful theory because Newton was a nice person."

..." No doubt, instances of cherry-picked and poorly-worded "gotcha" phrases will be pulled out of context. One example is worth mentioning quickly. Phil Jones in discussing the presentation of temperature reconstructions stated that 'I've just completed Mike's Nature trick of adding in the real temps to each series for the last 20 years (ie from 1981 onwards) and from 1961 for Keith's to hide the decline.'"

In other words, nothing to see here, boys will be boys, and the stuff that looks bad was taken out of context.

What was amazing is what happened next. For the first time in the history of Real Climate, Schmidt opened up the comments and let everyone have their say. He replied courteously to both triumphant skeptics and desolate global warming believers, continuing to give references and cite papers. It was a triumphant performance, and one that was badly needed by The Team. His calm demeanor and dogged insistence on the rightness of his cause truly calmed the blogosphere down until events passed out of the hands of bloggers and into the maw of the major media machine. In total, there were 2,022 comments submitted over the course of two or three days, and Schmidt passed into several record books for his blogging performance.

The scientists whose emails and files were released to the outside world on Wednesday have been shown to be climate bullies and people who tried to game the scientific system to make sure their Team always came out on top. But are there more serious consequences to their behavior?

So far, stories about the incident have focused on the unsavory behavior of the scientists forming the Team, and it's easy to see why. They apparently worked together to corrupt the peer review process so vital for the correct functioning of science, evade compliance with legitimate requests for information, hide defects in their data and to paper over an inconvenient lack of warming over the past decade.

Are there more serious consequences to what they've done? We think so, and we don't think we have far to look.

Real World Consequences of Climategate

Roger Pielke Sr. has had a long and distinguished research career, as associate professor at the University of Virginia, professor at Colorado State University, Duke and the University of Arizona, and has held a variety of responsible and interesting positions in and out of academia.

Pielke has spent the bulk of his career trying to advance the point of view that while CO_2 does contribute to climate change, other things that humans do have an equal impact on climate and temperatures. He thinks the most appropriate way to gauge what's happening to global temperatures is by measuring the heat of the oceans. He has written many scientific papers published in peer-reviewed journals advancing his position, noting that deforestation, conversion of land to agriculture, changes in land-use policy in general, depletion of ground water sources, all have impacts on the climate surrounding these areas.

Pielke's views are cogently expressed, coherent, backed up by numerous studies in the field--and to a considerable extent, they have been marginalized by critics from The Team, who viewed him as dangerous, to the extent that he could take the focus off their obsession, human emissions of CO_2. It didn't matter that Pielke believes (and has often written) that human emissions of CO_2 are a powerful factor. It didn't matter that he is as far from a skeptic as you can possibly be. It didn't matter that his academic credentials and publishing record rivaled, if not surpassed, their own.

The Team viewed Pielke (as they did so many others) as an enemy. The released emails contain references to him as an albatross, and tried to make him seem like one who had lost all credibility, and they were careful not to send him some details relevant to his work. He even featured in a crudely photoshopped cartoon, appearing alongside skeptics. In fact, because of his credentials, credibility and willingness to take anthropogenic global warming seriously, Pielke was more dangerous to The Team than any skeptic, much in the way a heretic is more dangerous to the priesthood than any atheist. Cheap and malicious behavior, replicated in the way The Team treated many others in the emails that we have seen.

But much as we might sympathise with Roger Pielke Sr., his treatment is secondary to the effect that the attempts to marginalize his position has had. If Pielke was right, and recent work is moving more towards his long-held position, then mitigation and adaptation to global warming could have begun a decade ago, could have had a material impact, could have been designed to have positive side effects on the people living in the areas affected, and all for a tiny fraction of the money being called for by the Big Gun approach to CO_2 that demands that we make drastic changes to our economies, tax structures and even legal systems.

It just didn't fit in with The Team's grand plans. So they cut him out, reduced him wherever possible to a bit player, and congratulated themselves on their clever work.

Roger Pielke is just one of dozens, maybe hundreds, of people who have been seriously trying to make positive contributions to the environmental health of this planet, including reducing global warming and its effects. How many other sound ideas have been overlooked or shot down because they didn't fit in with The Team's political agenda?

The Real Crime in Climategate

While some are checking the statute books regarding the different treatments of hackers versus whistleblowers, and others are checking conspiracy laws regarding damaging careers through perversion of the peer review process and suborning editors to exclude unpopular opinions, we would like to say what we think the real crime is in Climategate.

The criminals are not limited to The Team, the climate scientists and paleoclimatologists whose emails and files were leaked to the public.

A section of politically active scientists, policy makers, politicians and NGOs in effect put on white coats and told us that our planet was gravely ill, and that we needed to follow their prescriptive advice to save ourselves from a deadly disease. That's really how they framed the discussion, and they classified everyone who disagreed as a denier, like a smoker dismissing his cough and waving away the x-rays.

That's not a crime. But it's pretty close to it to change the readouts on a patient's condition to convince him to undergo expensive treatment, label other doctors as quacks if they disagree with the changed diagnosis, and to refuse to show the patient the data underlying the charts.

They may protest that the diagnosis is too technical for the patient to understand and that their actions are for the patient's good. They may even believe it. But we call it quackery.

And the crime is malpractice. Deliberate and conscious malpractice. And since they arrogated the power unto themselves to diagnose the disease and prescribe a cure, they might also be charged with practicing medicine without a license.

As we said at the beginning of this book, this battle between alarmists and skeptics is really about which vision of the world will guide us going forward. It's really important and it's a bit strange that leaked emails play such a big part in this. After all, the basic positions have been laid out for years, like trenches in a World War 1 battlefield.

One vision of the world holds the world at threat--menaced, once again, by man. The other holds man essentially a bit player, unable to really affect the natural processes that ebb and flow across the world's systems. If this sounds like taking their positions to extremes, well that's what both sides have done--to themselves and the other side.

Using science like a club is not really what science is meant for, but that's what's happening. The alarmists, organized and arrogant, trumpet study after study meant to show us that our emissions of CO_2 are wilting the world like an old head of lettuce. The major media outlets, long convinced the alarmists are right, find pretty pictures and anecdotes to illustrate the story. The skeptics, mostly disorganized (in more ways than one), keep finding flaws they think are fatal in the alarmist studies. The alarmists grow furious, which makes the skeptics more skeptical.

The leaked emails and documents show that the alarmist position is in part a facade, concealing shabby data handling and shabbier ethics--it's as if the data could be mailed in by a six-year-old because the issue was decided, and it was more entertaining to pore over the enemies list and decide which journal to boycott and which scientist to insult.

But what remains of the vision? If everything the skeptics allege about the wrongdoing pointed at (if not out) by the leaked emails is correct, is the alarmist vision dead? If so, does that mean the skeptic vision is triumphant?

As lukewarmers, people who believe in global warming but not that it will be catastrophic, we feel a bit of distance from both sides. We don't think that this kills the case for anthropogenic global warming. Indeed the emails reveal as much passion on the part of the scientists as it does pathos--part of their misbehavior seems rooted in the fierce conviction they were right, after all. But it may blow apart the most extreme predictions of large scale sea level rises and dramatic climbs in temperature. Which to our minds is all to the good.

Nor do we think that grounding the alarmists automatically wins the field for the skeptics, those who are convinced there is no warming, or that what warming there may be is 100% natural variability in action. What it does is make the case for more skepticism and more studies. It may be necessary to start some studies from scratch--maybe the skeptics can look on, or even participate in study design. Worse things could happen.

And maybe this episode makes a strong case that 'the vision thing' should not yet be a part of the study of human effects on climate change. That we should be looking down at the ground searching for evidence, not up to the skies for inspiration.

What we feel most strongly about right now is the removal of partisan politics from the issue. Having Republicans advancing skepticism and Democrats pushing alarmism is not going to be good for either party in the long run. But it will be worse for the two sides of the scientific coin-- and in the short run.

The next clarion call will be for those who wrote the code that crunched raw temperature data to explain the decisions they made for the adjustments to the numbers. We all know adjustments are necessary. But we will all want to know what changes were made and why. This will cause a repetition of the same arguments heard before the release of the emails and code--with skeptics cast as unscientific botherers by the alarmists, and the alarmists cast as intransigent data concealers by the skeptics. Unless we can find a different way to deal with each other.

Infant Science, Infantile Scientists

Because we posthumously bestow the title of scientist on greats of the past, we tend to forget that science as we know it today is pretty new. Although you can make the case that the scientific method is 500 years old, you could say with equal credibility that science as practiced today is just a bit over 150 years old. Despite the work of Svante Arrhenius at the end of the 19th Century, the study of climate change is a lot younger, dating only back to 1941 with the publication of Helmut Landsberg's book, Physical Climatology.

Certainly, many of the disciplines within science are young--young enough to be worth talking about when we discuss the latest of the infant sciences, climate science. And it's worth doing so, because when climate scientists insist on the correctness of their theories and accuracy of their observations, they're making quite a claim for themselves. Certainly no other new discipline was as accurate and far sighted as climate scientists say they are.

Grave error exists in new science, and crops up occasionally even in well-established fields. The suppression of Wegener's theories, the existence of Lysenkoism in the 1940s and 1950s, cold fusion, Piltdown Man, Lord Kelvin's estimate of the age of the Earth, all were very wrong--but looked at together, most can be classified as a 'natural' series of mistakes common to a new field of study. There are two exceptions in that list (which is by no means complete): Piltdown Man was a hoax, although it might be more apt to label it a prank. And the suppression of Wegener's theories was a mass rejection by the scientific establishment of an alternative theory that proved true—years after Wegener died.

Climate scientists could have used the newness of their discipline to plead for forbearance while they formed hypotheses and figured out the best ways to collect and measure data. They would have been given a lot of latitude had they tried. But political pressure forced them to claim invincibility and perfect accuracy--with a straight face. When hundreds of billions are staked on a scientific claim, there is no margin for error. At the end of the day, the scientists involved did not have the maturity to say no to wrongly aimed research grants, no to politicians' mangling of the science in public speeches, no to the twisting of their work to the point that they twisted with it. We see in the leaked emails that they acted like spoiled children much of the time, and they have a lot to answer for. So did those who spoiled them.

But errors there were, and the first scalp has been claimed with the temporary stepping down of Phil Jones as Director of the Climate Research Unit in East Anglia, pending an independent review of the Crutape Letters saga. No joy here--at the end of the day, if all The Team fall by the wayside, the really culpable parties--the politicians and pundits who pushed them to claim for their science what they could not in truth provide--will move on to the next topic, the next disaster, the next crisis demanding that we put more faith and invest more power in them. The politicians will not go away.

Global Warming and Cotton Candy

Taking a broad look at the history of global temperatures, the general story is very clear. At the end of the last ice age temperatures climbed very rapidly for 1,000 years, reaching a level plateau about 10,000 years ago. Since then temperatures have risen and fallen in cycles around a slowly increasing average. The period between 1975 and 1998 saw temperatures increase more quickly than anticipated, and more quickly than scientists thought would be normal. They thought that human contributions of CO_2 could explain this rapid increase and began searching for data to prove or disprove it.

Although many in the scientific community would call the above paragraph simplifed to the point of banality, we don't think anyone would disagree with it. Many will, however, disagree strongly with what follows.

Many individual studies designed to measure the contribution of man-made CO2 to temperature rises have ended up seeming insubstantial, with questions about methodology, data and analyses leaving onlookers wondering if there is any 'there' there. The net impression is of beautifully woven storylines that end up, on closer examination, as insubstantial as cotton candy.

The specific research undertaken to measure and analyze human contributions of CO2 to temperature rise were mostly flawed. This should not be surprising--it's an infant science and people were designing measurement strategies and tactics on the fly--and nobody should think the worse of scientists because of the mistakes that were made.

Instead of accusing scientists of bowing to a political agenda, we should perhaps look at them as independent agents (with strongly held beliefs about the subject) who were actually put under pressure to meet political deadlines for conferences and reports, and yet more pressure to make sure the story was clear and consistent--pressure from lay people who don't understand the ups and downs of science and the slow timeline to certainty.

What Steven Mosher calls The CRUtape Letters (the leaked emails of Climategate) provide an indication that some scientists may have folded under pressure and provided the showboat rising curves and data that the politicians needed so badly. I do not think it would be easy to say no to Al Gore. He looks like a bully. They probably looked at it as sort of shorthand--they believed that the final results of data examination would replicate these curves, so giving the politicians what they wanted was just jumping the gun a little.

But it came apart on closer examination. This normally would be part of the scientific process, and they would have grudgingly thanked their critics and gone back into the field under other circumstances. But the political pressure and time deadlines were too tough, so their critics became enemies, denialists and evil. And so science will suffer as a result. And while others are calling for more transparency and the openness of data, we will counter with a call for some mechanism to shield science and scientists from the pressure to produce the right results in the next 10 minutes or you'll lose your funding for the next 10 years.

We believe that global warming will increase average temperatures by about 1.5 to 2 degrees Celsius over the course of this century. But what will that mean in effect? Most of the predictions seem to come from people with a political stake in the outcome, and are hazy enough to not be of real help. We will tell you that in the course of writing this piece we were quickly and frequently reminded of how ignorant we are about many important things. But we confess also that we're a bit surprised at how many share our ignorance.

So far, all of the global warming we've been able to measure has occurred in the North, at night, and in winter. If the end result of global warming is to make Canadian and Russian winters more pleasant, few will mourn. But let's take a look at what is happening in the world and see what global warming might do.

Now, really, there is no actual average global temperature. Even if there were, how helpful would it be to know that a given temperature X was generated with temperatures from the Arctic and the Sahara? What useful decisions could you make from that? So let's look for other data that might help us gauge the effects of global warming.

The average annual temperature for Chicago is 9.5 degrees Celsius. The 10 fastest growing cities in the world have temperatures about 8-10 degrees Celsius higher than Chicago. So, I think we can say that if Chicago and Helsinki heat up, it may not cripple their growth prospects. But if the world's 10 fastest growing cities heat up 3 degrees more, what will it do to them? And this I do not know. But we're not really worried about the human animal--we already have large and thriving cities well beyond the extremes of temperatures that global warming will bring us. We'll spend some money, change a few habits and adapt. It's pretty much what we do. We're more concerned with the other things we need to help the human animal thrive--things like agricultural land, forests, a living ocean, adequate biodiversity.

For agriculture, one of the most productive agricultural regions in the world is California's Central Valley. Average winter temperatures there are 12.8 degrees, summers average 23.9. In Saskatchewan, the Canadian province that produces 45% of Canada's wheat, winter temperature averages range from -10 t0 -22 degrees, while summer temperatures range from 11 to 22 degrees. So Saskatchewan might not be hurt by moderate global warming. But what happens to California's Central Valley if their temperatures climb? Agriculture thrives in warmer and wetter climes, but what changes will need to be made to the crop mix? What different pests and plant diseases should we prepare for?

Global warming is supposed to bring more precipitation, although it won't be spread evenly across the globe. We assume it will still precipitate upon the unjust as well as the just... If all the world becomes like Blade Runner's Los Angeles, it will cause changes. If some regions become more like Phoenix, we will need a lot more nuclear power plants to run those air conditioners, and we will need to become rich enough to pay for them.

Which kind of brings us to our point. We don't know the answers to a lot of the questions raised above, and we didn't stumble across internet sites that credibly claimed to know the answers. But what we do know is that whatever changes global warming brings, it will be the poor of this planet who cannot move quickly enough, cannot change seeds for their croplands, cannot react to new pests, cannot weatherproof their homes, certainly cannot afford air conditioning, who will suffer most. It seems that for us, global warming may continue to have the air of a spectator sport, with Republicans and Democrats yelling at each other, and skeptics and alarmists producing swarms of internet links that prove each other wrong.

But if we cannot pinpoint the effects of global warming now--if we cannot tell the people of Bangladesh or the Congo what the near future holds--then it is incumbent upon us to make them strong enough that they can react to circumstances quickly and intelligently. They don't need us to be the Great White Fathers who stop Global Warming. They need us to be the friends who found out first how to exploit science and technology to give us the flexibility and mobility to cope with a changing world. The friends who share what they've learned and some of what they've earned.

Unless we want a diet of cotton candy to inform our future...

Global Warming and the Wolf

Being in the middle doesn't automatically make us right--but it may make it worthwhile when we call for both sides to pay attention to one thing. So both skeptics and supporters of AGW (anthropogenic global warming), we ask you to pay attention here.

The boy who cried wolf. Old fairy tale, in current vogue—you know the story. Aesop wrote of a bored shepherd boy who amused himself by yelling 'wolf' to draw the villagers to his unneeded rescue. Wolf really does show up, villagers don't, boy (and flock) gets eaten.

So, believers: Recently released was the umpteenth story about how global warming is getting worse, illustrated by very real Arctic melt and possible melting of Greenland, with the clear message that if temperatures continue to rise and Greenland's ice melts, sea levels will rise by a whopping six meters.

Wolf! The IPCC forecasts temperature rises, even under business as usual (no efforts to control emissions) up until 2100, after which they anticipate a decline, in line with falling world population and improved decarbonisation. But temperatures would have to rise until the year 3000 to melt the ice in Greenland, because most of the ice sits in a basin like ice cream in a bowl. So, yes, if the ice in Greenland melts, sea levels will rise six meters. But nobody, including the people who wrote the story, believes that's going to happen. They are writing things they know to be false with the express purpose of scaring you.

We could cite numerous other examples, and we're sure many will. But when believers wonder why support is declining for global warming policies, this is one of the reasons. Your leaders have cried wolf once too often. Part of this whole email scandal revolves around a Team of scientists finding more and more clever ways to keep crying wolf, changing the temperature numbers to make earlier temperatures look lower and recent temperatures look higher, so the change would look more dramatic. Wolf! They acted in a way that harmed public policy, you the public and worst of all, the trust we place in science.

Okay, skeptics. Your turn!

The wolf ate the boy, and the flock of sheep he was protecting. Get it? The wolf... ate the boy... and the flock of sheep he was protecting. There was a wolf.

Temperatures rose quickly between 1975 and 1998. Idiot alarmists focused on the top temperature reached and suppressed the historical data showing warmer periods in the past. But the troubling data is the speed with temperatures rose. We know that many things contributed to this rise: The natural recovery from the Little Ice Age, massive land use changes due to bringing enough land under the plow to handle the population explosion, deforestation, desertification, etc. But CO_2 was one of the contributors. We don't to this day know how much of a contributor, but it would be mind-bogglingly stupid to think it played only a trivial part. We went from no cars a century ago to many, many cars now. We went from zero power generating plants a century ago to lots and lots of them now. (Like our precise statistics?)

Warming might be beneficial to humans--certainly to some of them. But if CO_2 is a major contributor to increased warming, the next wave will come elsewhere. It plausibly could turn some developing countries into deserts with no agriculture, and others into flood-and-drain combination hellholes where agriculture is the last thing on peoples' minds. Because global warming really doesn't exist. It's expressed as regional warming, and it's unpredictable where it may strike and how quickly temperatures will rise. We said that we believe global average temperatures will rise about 1.5 to 2 degrees Celsius. But it won't be even, and it won't happen smoothly over the rest of the century. It will hit some places like a ton of bricks and leave others untouched. A slow motion tornado that picks and chooses.

The idiots running the global warming campaigns (and make no mistake, a bigger bunch of idiots would be hard to find) didn't trust people to react to the truth. They thought you wouldn't understand and if you understood that you wouldn't care. So they lied to you, repeatedly and with a smile on their smarmy faces. Catastrophe! Dramatic sea level rises! Unbearable heat waves every x months! Pick your own stupid pet trick.

One of the consequences of the shift in climate science towards dependence on paleoclimatology and temperature measurements has been an inevitable shift away from the direct observations of the real world. Although current climate scientists certainly pay lip service to melting glaciers, migration patterns of birds and beetles and the precession of the seasons, the agenda is driven by numbers on charts. And those numbers are processed.

If we were depending on the natural observations of glaciers, birds and beetles and early springs, there would probably not be an IPCC, or a summit meeting in Copenhagen. Although news headlines find more and more instances of these phenomena, they are not new, and skeptics, instead of poring over cryptic emails, would just be citing instances of similar events from the past.

But as clearly as we are able to see the truth, we have written it here. Global warming is real and it is a problem, if not the catastrophe they want you to believe. It needs our attention. CO_2 is a contributor, along with other factors, some natural and some manmade. And yes, we do need to do something about it.

CPSIA information can be obtained
at www.ICGtesting.com
Printed in the USA
BVHW020034050123
655621BV00019B/159

9 781450 512435